Disability and Work

Disability and Work

Incentives, Rights, and Opportunities

Carolyn L. Weaver, editor

The AEI Press

Publisher for the American Enterprise Institute
WASHINGTON, D.C.

1991

Distributed by arrangement with

University Press of America, Inc.
4720 Boston Way   3 Henrietta Street
Lanham, MD 20706   London WC2E 8LU England

Library of Congress Cataloging-in-Publication Data

Disability and work : incentives, rights, and opportunities / Carolyn
   L. Weaver, editor.
      p.   cm.
   Includes bibliographical references.
   ISBN 0-8447-3738-0
   1. Handicapped—Employment—United States.
   2. Handicapped—Employment—Government policy—
United States. I. Weaver, Carolyn L.
   HD7256.U5D58   1991
   331.5'9'0973—dc20   90-24886
                                                          CIP

1 3 5 7 9 10 8 6 4 2

AEI Studies 516

Printed in the United States of America

Contents

Preface

On July 26, 1990, President Bush signed into law the Americans with Disabilities Act.[1] Widely regarded as landmark legislation, the ADA extends civil rights protections to the disabled and mandates that private businesses modify jobs, facilities, programs, and policies to ensure them "equal access" and "equal opportunity." In addition, it mandates that telecommunications services must be accessible to the hearing- and speech-impaired and that public transportation must be accessible to the disabled generally. Proponents of this sweeping legislation estimate that there are some 43 million people with disabilities who will benefit from the ADA. Virtually all American businesses will be affected.

This book takes a critical look at the employment provisions of the ADA. How will the law affect the wages and employment of disabled people—and at what price to American businesses and consumers? How will other workers be affected, including those protected by other civil rights laws? What are the likely effects on labor costs and product prices, and on the efficiency and competitiveness of American firms? The costs and benefits of this highly regulatory approach to promoting work among the disabled are contrasted with those of incentive-based approaches.

The book also explores the problems of defining and deciding who is disabled for purposes of distributing government benefits, services, and guarantees. According to Evan Kemp, chairman, Equal Employment Opportunity Commission, in this volume, the federal government uses over forty definitions of disability and state and local governments use literally thousands more. The question arises of whether it is possible to decide, in a systematic and cost-effective manner, who is or is not disabled for work. What are the costs and consequences of making wrong decisions? Valuable insights into these problems are gleaned from our experience with the social security disability insurance program.

As the chapters in this book highlight, disability is not a concrete, black-or-white concept like race or sex. There are over a thousand distinct physical and mental impairments—including diseases, injuries, disorders, illnesses, deformities, and other abnormalities—which range from completely debilitating to slight in their impact on work ability.[2] How disabling an impairment will be depends on a wide variety of individual factors, including education, income, motivation, job opportunities, age, medical treatment, vocational rehabilitation, and family and community support. An impairment with relatively minor impacts on the labor market activity of one individual can be totally incapacitating, either temporarily or permanently, for another.

In other words, disability is not readily observable, and it is not a static condition easily classified as present or absent for a lifetime. It can be affected in important ways by the incentives and disincentives inherent in government policy. This imprecision creates difficult problems for the design and control of public programs for the disabled.

This book seeks to improve our understanding of the complex interrelationships between physical or mental impairments, disability, work, and government policy—an understanding that is critical for the effective reform of the nation's policies for the disabled. It reflects a broad range of views from a distinguished group of scholars and public officials interested in disability policy. The contributors draw on a solid base of economic and legal research in the areas of disability and labor force participation, labor market discrimination, economic regulation, and administrative law.

The chapters in this book are based on papers and remarks delivered at a policy forum, "Disability and Work: Incentives, Rights, and Opportunities," held at the American Enterprise Institute in September 1989. Funding for the policy forum was provided by the Pew Charitable Trusts.

CAROLYN L. WEAVER

Contributors

CAROLYN L. WEAVER, editor, is resident scholar at the American Enterprise Institute and director of AEI's Social Security and Pension Project. She has been a senior research fellow at the Hoover Institution and a member of the economics faculties of Virginia Polytechnic Institute and State University and Tulane University. From 1981 to 1984 she served on the staff of the U.S. Senate Finance Committee, with primary responsibility for social security and disability legislation. Ms. Weaver was a member of the 1987 Disability Advisory Council and the 1989 Disability Advisory Committee. She was a contributor to *Disability and the Labor Market: Economic Problems, Policies, and Programs*, edited by Monroe Berkowitz and M. Anne Hill (1986), and has written extensively on social security and political economy.

MICHAEL J. ASTRUE is general counsel of the Department of Health and Human Services. Previously he served as associate counsel to the president and as counselor to the commissioner of social security and as acting deputy assistant secretary for human services legislation. Mr. Astrue received his law degree from Harvard Law School and is a member of the Massachusetts bar.

RICHARD V. BURKHAUSER is professor of economics and senior fellow in the Gerontology Center, Syracuse University. He has published widely on the behavioral and income-distribution effects of government policy toward aged and disabled persons. He is coauthor of *Disability and Work: The Economics of American Policy* (1982) and of *Public Policy toward Disabled Workers: A Cross-National Analysis of Economic Impacts* (1984). His latest work on this topic appeared in the *Milbank Quarterly* (1989).

MARTIN H. GERRY is assistant secretary for planning and evaluation at the Department of Health and Human Services. He is a former director of the department's Office for Civil Rights. Before his appointment at HHS he was a visiting scholar at Stanford University, special counsel to the Wednesday Group of the House of Representatives, president of the Policy Center for Children and Youth, and a member of the 1989 Disability Advisory Committee of the Social Security Administration. Mr. Gerry received his law degree from Stanford Law School and is a member of the bars of Washington, D.C., the state of New York, and the U.S. Supreme Court.

EVAN J. KEMP, JR., is chairman of the U.S. Equal Employment Opportunity Commission. He has written numerous articles and spoken extensively about issues that concern disabled people and, for the last seven years, has taught a course on disability and law at Catholic University Law School. From 1980 to 1986 he was head of the Disability Rights Center. Mr. Kemp received his law degree from the University of Virginia.

JONATHAN S. LEONARD holds the Harold Furst Chair in Management Philosophy and Values at the Haas School of Business, University of California at Berkeley. He is also research associate of the National Bureau of Economic Research and the Institute of Industrial Relations. Mr. Leonard recently served as senior economist on the President's Council of Economic Advisers, and he has served as an adviser to numerous government agencies. He has published research papers on welfare compensation, job creation, affirmative action, industrial relations, unemployment, and disability policy and is coeditor of the *Journal of Human Resources* and of *Industrial Relations*.

JERRY L. MASHAW is the William Nelson Cromwell Professor of Law and a professor at the Institute for Social and Policy Studies, Yale University. He has also served on the law faculties of Tulane University and the University of Virginia, and he has written numerous books and articles on administrative law, regulation, and social welfare policy. He is the author of *Bureaucratic Justice: Managing Social Security Disability Claims* (1983). Mr. Mashaw is a member of the American Academy of Social Insurance and an occasional consultant to various government agencies and private foundations.

WALTER Y. OI is the Elmer B. Milliman Professor of Economics at the University of Rochester. He has held positions at Northwestern University and the University of Washington and has been a visiting scholar at the Hoover Institution and at Princeton University. Mr. Oi is a fellow of the Econometrics Society and president–elect of the Western Economic Association. He has published widely in the fields of labor economics, price theory, and applied econometrics.

DONALD O. PARSONS is professor of economics at Ohio State University. He has published extensively on the interrelationship between health, social insurance, and labor market activity, including articles in the *American Economic Review*, the *Journal of Political Economy*, and *Economica*. His other research interests include employment contracting and job mobility, intergenerational wealth transfers, and minimum wage policy.

SHERWIN ROSEN is the Bergman Professor of Economics and chairman of the Department of Economics at the University of Chicago. He is also research associate at the National Bureau of Economic Research and at the Hoover Institution, fellow of the American Academy of Arts and Sciences and of the Econometric Society, and editor of the *Journal of Political Economy*. Mr. Rosen has published widely in the areas of labor economics, price theory, and the theory of contracting and optimal incentive structures.

Disability and Work

PART ONE

Incentives-based or Rights-based Approaches to Promoting Employment?

1
Incentives versus Controls in Federal Disability Policy

Carolyn L. Weaver

The rising cost of federal disability programs, along with a heightened awareness of the ability of severely disabled people to work, has fueled a rethinking of government policy in recent years.[1] Since at least the mid-1970s attention has shifted from income support for the disabled to policies designed to promote independence, freedom of choice, and, where possible, employment. In the area of employment, the focus of attention has been on improving incentives to work.

A series of reforms has been adopted in the two largest cash programs for the disabled, social security disability insurance (DI) and supplemental security income (SSI), to reorient these programs away from their original conception as income support for "early retirees."[2] Explicit work incentives and improved delivery of rehabilitation services have been key aspects of this reorientation. In 1980, for example, Congress revamped the way SSI benefits were paid to the disabled poor so that these individuals could work and continue to receive partial cash payments and Medicaid. Previously, individuals were judged to be either capable or incapable of work, and thus either eligible or ineligible for benefits. In the DI program Medicare was made available to beneficiaries for four years after they returned to work. Individuals who subsequently returned to the benefit rolls were reentitled to Medicare without having to meet the typical two-year waiting period. In addition, individuals were permitted to receive cash benefits for a year after returning to work and could be reinstated to benefit status within the following two years without having to meet the typical five-month waiting period. To help ensure that benefits did not exceed predisability earnings, the maximum family benefit provisions were tightened, as was the method of computing average earnings for younger workers. In the area of

3

rehabilitation, Congress changed the way it financed state rehabilitation agencies serving DI and SSI recipients; resources were targeted toward those successful at getting recipients back to work. Though modest in scope, these and the other work incentive provisions adopted in 1980 were unparalleled in the history of the social security disability programs.[3]

The work incentive provisions in DI and SSI were complemented by a set of tax provisions adopted in the mid- to late 1970s. These included the targeted jobs tax credit, which allows employers a credit for hiring people with disabilities; section 190 of the Internal Revenue Code, which allows employers a deduction for expenses incurred in eliminating architectural and transportation barriers; and more generally, the earned income tax credit, which provides a refundable tax credit against earned income for all low-income people with children.[4]

The common problem being addressed by these seemingly diverse reforms was that the design of government programs was inhibiting rather than encouraging work. The experience with DI during the 1970s gave ample reason for concern. Between 1970 and 1978, the number of workers on the DI rolls nearly doubled, from 1.5 million to 2.9 million, and expenditures quadrupled. By 1980, some 4.7 million Americans, including workers and their dependents, were receiving cash benefits at a cost of $16 billion annually. In the span of just a decade, the cost of the program for the average earner increased by more than 50 percent.[5] Over roughly the same time period, the number of people returning to work fell to an all-time low, despite a steady decline in the average age of new beneficiaries. In 1976, somewhat fewer than 1.5 percent of beneficiaries were leaving the rolls each year to work—fewer than the number who died or converted to retirement benefit status.[6] (The figure is even lower today.)

The various reforms adopted in 1980 were designed to reduce the relative attractiveness of DI for those still at work, and, for those on the rolls, to increase the skills they brought to the labor market and increase the relative attractiveness (or reduce the relative risks) of returning to work. They embodied an incentive-based approach to the problem of promoting the employment of the disabled.

A Shift toward Controls

In 1990, the incentive-based approach was abandoned for a rights-based approach. The Americans with Disabilities Act is a comprehensive civil rights bill that bans discrimination in the public and private sectors in employment, public accommodations, public services, transportation, and telecommunications.[7] Of immediate interest are the employment provisions, which create new rights for the disabled and impose new rules and regulations on firms, backed up by the enforcement authority of the U.S. Equal Employment Opportunity Commission (EEOC) and the federal courts. The problem of the employment of the disabled is approached from the demand side of the labor market—that is, as a failure of employers to hire, retain, or promote qualified workers with disabilities on equal terms with other workers—that must be remedied through mandates rather than incentives.

The underlying premise of the new law is that large numbers of disabled people are willing and able to work but are being denied the opportunity—either by outright and widespread discrimination or by the failure of employers to make reasonable adjustments in jobs or physical surroundings. According to proponents, employers are preventing the disabled from achieving their full potential as productive, taxpaying citizens; both the economy and the federal budget stand to gain from a policy that moves these "second-class citizens" into the "mainstream" of the American economy.[8]

Because critical aspects of the ADA have been left to federal regulators and judges to define and to interpret, it is not possible to predict with certainty what the effect of the law will be on the well-being of the disabled and of citizens generally. Certainly the law will increase employment and wages, perhaps quite substantially, for some disabled people. But for many others, particularly low-skilled, severely impaired individuals, the gains are likely to be modest at best and could well be negative. The potential clearly exists, however, for the gains to the disabled to be outweighed by large economic costs to society as a whole.

The central flaw of the ADA is not in the extension of "the same civil rights protections to people with disabilities that now apply to people on the basis of race, sex, religion, and national origin," to use the language of the bill's sponsors. Rather it is in the imposition on employers of a duty to "accommodate" the mental or physical limitations of the disabled worker or applicant without weighing the expected benefits of such accommodation. This, in combination with the antidiscrimination provision, distorts a civil rights measure into what is essentially a mandated benefits program for the disabled.

The Meaning of Discrimination

To economists, discrimination means failing to hire, retain, promote, or compensate identical people on equal terms, where identical means making the same net contribution to a firm's output or profits.[9] For people with disabilities, discrimination might involve separate or inferior treatment resulting from, say, a repulsion toward severe deformities, from unfounded fears of (noncontagious) diseases, or from misjudgments about the mental capabilities or social skills of

5

people with physical impairments—examples commonly cited by proponents of the ADA.

To have an adverse effect on the wages and employment of disabled people, discrimination must be manifest in an employer's willingness to incur some cost to avoid disabled people. To see this, consider a firm that discriminates against blacks. From an economic perspective, this means that the firm is unwilling to hire blacks at the same wage (or total compensation) as equally productive whites, or, stated another way, it demands fewer blacks than whites at any particular wage. The reduced demand for blacks among firms that discriminate reduces the relative wages of blacks, creating an economic opportunity for nondiscriminators: firms that do not discriminate can hire these equally productive workers and produce the same output at lower economic cost. Lost profits are the price an "economic discriminator" pays to indulge his preferences.[10]

The imposition of an antidiscrimination law (of the equal-pay-for-equal-work sort) implicitly taxes this preference for discrimination, increasing the demand for minority workers and raising their wages. Productivity need not suffer since blacks are perfect substitutes for equally qualified whites, in the sense that they make the same net contribution to the firm's output or profits.[11]

As disabled rights groups are keenly aware, the ADA goes well beyond this type of equal-pay-for-equal-work requirement. Rather than amend civil rights statutes to include people with disabilities, the law creates a new civil rights statute that mandates unequal treatment of equals. It begins with the deceptively simple statement that

> no covered entity shall discriminate against a qualified individual with a disability because of the disability of such individual in regard to job application procedures, the hiring, advancement, or discharge of employees, employee compensation, job training, and other terms, conditions, and privileges of employment.[12]

It then goes on to define a "qualified individual" as, first, disabled within the meaning of the law, and second, able to perform the "essential functions" of the job *with or without reasonable accommodation.*[13] The provision applies to all firms with fifteen or more employees. The remedies and procedures of title 7 of the Civil Rights Act of 1964 apply.[14]

The legislation thus includes in the protected population people who, in an economic sense, are not as productive or do not make the same contribution to the profitability of the firm as other people with the same qualifications. (These are the people who can perform only

6

the essential functions of the job and who can do so only with accommodation.) While promoting the employment of this much broader group may be a highly desirable social goal, the antidiscrimination–reasonable accommodation approach is a costly and inefficient way of doing so and is likely to have highly undesirable distributional consequences.

What Is "Reasonable Accommodation"?

In defining discrimination and reasonable accommodation, the ADA draws on the statutory language and implementing regulations of the Rehabilitation Act of 1973. Sections 501, 503, and 504 of the rehabilitation act bar discrimination against the disabled by the federal government, federal contractors, and recipients of federal funds.[15]

Under both the rehabilitation act and the ADA, firms are required to make the accommodations to a job or workplace necessary to permit a "qualified individual with a disability" to perform the "essential functions" of the job the individual desires or holds.[16] Such accommodations are defined to include (but are not limited to):

- making existing facilities readily accessible to and usable by individuals with disabilities
- job restructuring
- part-time or modified work schedules
- reassignment to a vacant position
- acquisition or modification of equipment or devices
- adjustment or modification of examinations, training materials, or policies
- the provision of qualified readers or interpreters, and other similar accommodations[17]

A firm can escape this obligation only if it can demonstrate that the accommodation would impose an "undue hardship" on its operations, which is defined as "an action requiring significant difficulty or expense." Factors to be considered in assessing hardship include the size and financial resources of the business, the type of business, the composition of the work force, and the cost and nature of the accommodation.[18]

If the regulations implementing the ADA follow those of the rehabilitation act, no guidance will be provided to employers about how to weigh these factors; the only thing that will be spelled out— and only by example in an appendix to the regulations—is that larger enterprises will be expected to incur larger costs. The appendix to the section 504 regulations states that a small day-care center might not be required to expend more than a nominal sum, such as that necessary to equip a telephone for use by a secretary with impaired hearing, but a large school district might be required to make available a teacher's aide to a blind applicant. Further, it might be considered reasonable to require a state welfare agency to accommodate a deaf employee by providing an interpreter, while it would constitute an undue hardship to impose that requirement on a provider of foster home care services.[19]

From an economic perspective, two features of the reasonable accommodation requirement stand out: first, the absence of any reference to expected benefits in the determination of the costs that must be incurred on behalf of the protected individual; and second, the uncertainty of the legal standard. Before assessing the likely effects of the legislation, it is first necessary to consider the determinants of accommodation in private, unregulated markets.

Accommodation in Unregulated Markets

In private, unregulated markets, the decision to hire a disabled person and to incur costs in doing so is influenced by the same set of factors that influence the decision to hire a "nondisabled" person. There are costs associated with recruiting, screening, hiring, training, and otherwise accommodating different individuals. These costs can be fixed or can vary with the firm's output, and they can differ from firm to firm, from occupation to occupation, and even from person to person. They can be quite nominal—as in the case of a laborer on a small construction job, or a hearing-impaired person in an editorial position—or quite substantial—as in the case of a manager or executive at IBM, or a deaf person in a teaching position. These costs are present for economic discriminators (employers who fail to pay on the basis of net contribution to output or profits) and nondiscriminators alike. Among discriminators, these costs, in combination with wages, are simply evaluated to be higher when contemplating hiring a disabled person over a "nondisabled" person.

Wheelchair ramps for the disabled, job coaches for the mentally retarded, and interpreters for the deaf—no less than pensions, on-the-job-training, and day care—will be provided in private markets if they can be structured as part of mutually beneficial, productivity-enhancing arrangements between workers and employers. For example, an employer may well furnish and bear a substantial portion of the cost of a wheelchair ramp and other physical accommodations for a disabled employee if, as a result, accessibility is enhanced for disabled customers and future disabled employees. An interpreter for the deaf, on the other hand, which provides quite specific benefits to the deaf worker, is likely to be offered only if the worker judges the expense to be worth incurring (in the form of reduced wages).[20]

The size or profitability of the firm bears no necessary relationship to the decision to make or not make investments of this kind. The critical determinant is the relationship between expected benefits and expected costs, which determines expected returns. In competitive markets, firms cannot incur the expense of an accommodation (or pension plan or on-the-job-training program) unless they expect to recoup enough extra output or sales in the future to make the investment worthwhile. If labor is mobile, workers will not incur the cost unless they expect to recoup it through either improved employment opportunities or increased wages later. If both the worker and the employer expect positive returns on an investment in accommodations, we would expect the employer to offer it, the employee to accept it, and both to be made better off.[21] It is the flexibility of wages and other components of the compensation package that allows the interests of employees and employers to be aligned to generate these positive market outcomes.[22]

By directly regulating the terms of employment contracts with disabled people—banning certain arrangements and dictating the terms of others—the antidiscrimination–reasonable accommodation provisions of the ADA will increase the cost of hiring disabled people and distort business decisions. The law mandates accommodation but generally precludes adjustments in wages or other compensation; it mandates a level of accommodation without regard to expected benefits; and it does so for economic discriminators and nondiscriminators alike. The law therefore is not structured to economize on the resources needed to achieve wage and employment gains for the disabled. The extent of the inefficiencies (and the wealth transfers that will inevitably take place) will be determined largely by the way in which the EEOC and the federal courts interpret the quite imprecise rights and responsibilities of individuals under the new law.

Uncertainty of the Legal Standard

Another key problem with the ADA is that it enables a highly uncertain legal standard, which will tend to undermine decision making. The definition of reasonable accommodation, for example, adopted from the rehabilitation act, has generated considerable litigation, and to date, the case law has defined only the very broadest parameters of the employer's duty to accommodate. As the U.S. Civil

Rights Commission reported in 1983, "Doctrines governing the duty to provide reasonable accommodation are still in a formative stage. Neither judicial decisions nor regulations interpreting handicap discrimination laws clearly define this key legal concept."[23]

It is important to realize that the employer's obligation to accommodate and the determination of what type of accommodation should be made must be evaluated on a case-by-case basis, taking into account limitations of the individual, the nature of the job, and the financial and other circumstances of the firm. What is reasonable for one employer need not be reasonable for another; an individual who is qualified for a particular job and protected under the bill at one firm need not be protected at another firm.

Moreover, the determination of whether an individual is disabled, what accommodations are necessary or appropriate, and whether the individual, with these accommodations, can perform the essential functions of the job, must be made *before* medical evidence is gathered or inquiries made about the nature or severity of the individual's impairment. Under the legislation, such investigations— considered inherently discriminatory—must await the awarding of a job offer.[24]

Under the ADA, an individual is disabled if he or she has a physical or mental impairment that "substantially limits one or more of the major life activities," has a record of such an impairment, or is regarded as having such an impairment.[25] This means that the individual need not be currently disabled or limited in his ability to work in order to qualify for full protection under the law. If the regulations follow those for the rehabilitation act, moreover, there will be no clarification of the meaning of "substantial limits," and the list of major life activities will be suggestive rather than inclusive.[26]

If history is any guide, there will be no federal guidelines to define for employers the essential versus the inessential functions of different jobs or occupations, the types of accommodations that are appropriate for different disabilities of jobs, or even the range and severity of impairments covered by the legislation. For all but the most severely impaired or those with obvious physical or mental impairments, the question of who is disabled, and thus who may file a claim of civil rights violations against an employer, will be unclear.

These uncertainties will be compounded by practical questions about what constitutes discrimination, questions that do not arise in civil rights cases involving other minorities. For example, is it illegal to hire one disabled person rather than another if he (the former) requires less costly accommodation? Is it illegal to hire an individual (disabled or not) who can perform all aspects of a job rather than a disabled person who can perform only the essential aspects? If it is illegal to reduce wages to share the cost of interpreters or phone adapters, what about the cost of reduced work schedules, reduced workloads, or reduced speed and accuracy? Can speed and accuracy be considered essential functions of a job and thus grounds for different treatment? When is it legal to lay off or deny employment to an individual with a communicable disease, such as AIDS?[27]

Firms will respond to these uncertainties in different ways. Some will overcomply; some will undercomply. Too much accommodation—that is, accommodation that costs society more than it benefits it—may result, exacerbating the economic inefficiencies caused by the reasonable accommodation mandate itself. Litigation is likely to be the order of the day.

Altering the Incentives of Employees and Employers

What then are the likely effects of the employment provisions of the ADA? For purposes of discussion, it will be useful to distinguish among four elements of the legislation: reasonable accommodation, equal pay, equal employment opportunity, and remedies and sanctions.

The reasonable accommodation and equal pay provisions, in isolation, raise the cost of hiring workers with disabilities; without the equal employment opportunity requirement (and the sanctions), they would reduce the demand for disabled workers and consequently the number of disabled people employed. This increase in the cost of hiring people with disabilities, moreover, will not be commensurate with the workers' added value. As a result, the reasonable accommodation and equal pay provisions will distort the demand for workers with different kinds of disabilities and alter the relative wages of different groups.

The ADA will give firms the incentive to hire

- Individuals with relatively stable medical conditions, where the firm's exposure to liability is capped or can be reasonably assessed. This would suggest a bias toward individuals who are, for example, blind or deaf or amputees or paraplegics (without other complications), in preference to individuals with degenerative diseases, such as Parkinson's disease, or diseases with episodic symptoms, such as certain forms of mental illness.
- Individuals requiring accommodations that have spillover benefits to the firm, such as accommodations that improve accessibility for disabled customers or help the firm comply with the public accommodations provisions of the law. This would suggest a bias toward people who are mobility-impaired.

11

• Individuals requiring no accommodation or relatively inexpensive accommodation, either because of the nature or the severity of the impairment, or the individual's own contribution to accommodation (such as education or retraining), or the firm's prior investments in accommodation.[28]

• Individuals with more work experience, about whom the firm has a greater degree of confidence regarding likely performance and retention (and thus can make better assessments of the risk of future litigation).

The employment provisions will also influence the decision to retain disabled employees or hire new ones. Firms will have the incentive to retain disabled workers during periods of slack demand rather than lay them off and rehire others later, given their relatively greater knowledge they will have about the capability, reliability, needs, and potential liability of present employees.

Furthermore, the employment provisions will influence decisions about hiring people who are not disabled. Firms will have the incentive to hire individuals who are least likely to become disabled and thus to impose an unexpected liability. This would suggest a bias toward younger, higher-income, better-educated workers in preference to older, lower-income, more poorly educated ones. In addition, firms will tend to substitute toward disabled workers and away from less effectively protected minorities or classes (such as white women), and people not directly protected by civil rights statutes (white men).

Finally, as firms strive to keep production costs to a minimum and remain competitive, they are likely to reduce certain types of compensation that escape the scrutiny of the EEOC or fall outside the scope of the legislation. This might include informal or ad hoc on-the-job-training programs, special housing or transportation programs for unskilled workers, or job counseling and medical treatment for employees who have just become disabled. More formal programs that apply to all disabled workers, or to disabled and nondisabled workers of varying skill levels, may have to be reduced in scale or discontinued altogether. Keeping the cost of disabled workers—and labor generally—in line with their net contribution to the firm's output will be the motivating force. Unskilled workers in entry level jobs are likely to bear the brunt of these adjustments.

While many disabled people will benefit from what amounts to a program of employment subsidies, a large share of them will find the adjustments that result less than optimal. Some disabled people will receive too much of their compensation in the form of accommodation, and some will be deprived of job opportunities altogether; all the while, some will continue to be accommodated inadequately by economic discriminators.[29] Since disabled people will represent a potential legal liability for employers, their ability to negotiate directly with individual employers for preferable compensation packages (involving, say, more training or counseling and less accommodation, or less total compensation for a job) will be severely limited.

The ADA is also likely to alter the incentives of people with disabilities. In an unregulated market, individuals have an incentive to search for the best compensation package they can find. For disabled workers, this means searching for employers who are willing and able to accommodate their disabilities with the smallest adjustment in wages and other compensation. The sorting of disabled workers among firms as they search for the lowest-cost accommodators enhances efficiency. While the ADA will not eliminate this incentive (surely there will remain some adjustments that firms can make), it will dull it considerably. The search for employment by people with disabilities will now be biased toward those firms apparently most obliged to make costly accommodations—larger, more profitable firms—which may or may not be the ones best suited to accommodate their disabilities. As before, too many resources will be devoted to accommodation.

Similarly, the ADA will reduce the incentive of disabled people, rich and poor alike, to share in the costs of accommodation. Without the price mechanism of unregulated markets, which tends to align the interests of workers and employers, disabled employees will be unlikely to deem the same level of accommodation "reasonable" as do employers. The range of conflict and the likelihood of litigation can be expected to increase.

The Costs and Benefits

By increasing the cost of hiring disabled people, the reasonable accommodation and equal pay requirements, in isolation, will discourage the employment of the disabled. The more broadly the EEOC, through its rulings and regulations, and the federal courts, through their judicial decisions, interpret the responsibilities of firms, the greater the likely costs and adverse employment effects.

The ADA can have its hoped-for effects on the wages and employment of disabled people only if it is aggressively enforced. The degree of enforcement will be determined by the willingness of aggrieved individuals to press their claims before the EEOC and, if necessary, the courts, and by the willingness of the EEOC to redirect its limited resources from complaints involving other minorities toward those involving the disabled. With aggressive enforcement, the employment and wages of many disabled workers will surely rise.

13

Paradoxically, though, the extent of these gains is likely to be in direct proportion to the costs (the economic inefficiencies) that the courts are willing to impose on the private sector. As the courts decide just who is covered and just what is reasonable, they have the capacity to alter fundamentally the costs and benefits of the legislation.

With these caveats stated, let us examine the probable impact of the legislation.[30] In the case of the many disabled people covered by the bill who need no accommodation at all in the jobs they hold or seek, the reasonable accommodation provision should impose relatively small inefficiencies. (This group comprises people who, for example, have had cancer that is in remission, who are recovering alcoholics, or who walk with a limp.) In addition, the equal pay and equal employment opportunity provisions should serve to root out true economic discrimination. The primary inefficiencies that result from the ADA will be caused by, first, a substitution away from individuals who can perform all aspects of their jobs toward disabled individuals who cannot, and second, a general rise in the cost of laying off disabled workers.[31] This latter inefficiency tends to accompany any civil rights legislation; people who already have jobs benefit at the expense of those who do not.

Many other people covered by the legislation will need relatively inexpensive accommodations, some of which will have been provided privately through agreements with employers. To the extent these agreements were reached voluntarily, they tend to reflect an efficient level of accommodation in conjunction with wages and other compensation. In these cases, the legislation will generally force firms to increase wages and quite possibly to invest too much in accommodation, or both. In firms that are economic discriminators, the legislation will likely increase the wages and employment of disabled people; because of the absence of any cost-benefit evaluation, however, the adjustments could be either inadequate or excessive.

Still others covered by the legislation will require very costly accommodations and, as before, some firms will be required to spend too much on accommodation, wages, or both for some portion of them. As in the case of people needing lower-cost accommodations, the gauge of excessive costs is not the cost of the accommodation in relation to profits; it is the cost in relation to the expected benefits.[32] If expected costs exceed expected benefits, then the resources devoted to other uses—whether plant and equipment, research and development, other private investments, or hiring and accommodating other disabled workers—would have yielded higher returns. The cost to society in resources forfeited would exceed the gain to the affected disabled people.

14

In the final analysis, the civil rights–reasonable accommodation approach is not only costly, but also inherently limited in its ability to help the nation's disabled. Many disabled people, after all, are too severely impaired to work, given their age and education, and many others have impairments of such severity or complexity that, given the state of our knowledge and technology, they are not suited for competitive employment. For them, the ADA amounts to an entitlement to something they cannot use. Still others have impairments sufficiently severe that, given the availability of DI, SSI, Medicare, and Medicaid, they simply are not in the market for employment. For them, the ADA amounts to an entitlement to something they will not use—at least not until the value of work relative to income support is altered sufficiently. We simply do not know how large the remaining group is—those currently employed or with reasonable expectations of becoming employed with new accommodations—for whom potentially large economic costs will be incurred.

Reasonable Accommodation as a Mandated Benefit

A good analogy for the antidiscrimination–reasonable accommodation provisions in the ADA is a mandated benefit program for the disabled. The goal of a mandated benefit program—whether it be health benefits, maternal leave, higher pay, or more accommodation—might be a high priority from the standpoint of national policy, yet a rights- or entitlement-based approach is a wrongheaded way of achieving it. Such a program amounts to an off-budget spending program that increases labor costs, inhibits job formation, and produces an undesirable distribution of gainers and losers.

In the case of the ADA, income and wealth will tend to be transferred from consumers generally (who will bear the real cost of the mandate through higher product prices) and "nondisabled" substitute workers (who will bear the cost through a decreased demand for their services) to a segment of the disabled population, primarily the more highly skilled who already have jobs. The opportunities of some disabled people, particularly the unskilled, severely impaired whose opportunities are already restricted by the minimum wage, may well be further restricted.

The Uniqueness of "Disability" as a Minority Classification

The civil rights approach is limited in its ability to bestow benefits on the disabled without imposing significant economic costs, and the reason lies in no small measure in its failure to recognize the unique nature of disability.[33] Unlike race or sex, disability (or the underlying physical or mental impairment) generally places a limitation on work ability—either by reducing productivity on a particular job or by limiting the jobs the individual is capable of performing.

In addition, unlike race or sex, disability is a complex and changing phenomenon. An individual may be born disabled or may become disabled. The underlying physical or mental impairment may deteriorate or improve, may disappear or reappear, or may lead to death. It may have an impact on work ability now and not later, or later and not now.

Finally, unlike race or sex, disability is generally affected by individual choices. Decisions regarding occupation (coal mining, construction, or college teaching), lifestyle (smoking, drinking, drug use, or skydiving), education, training, rehabilitation, insurance coverage, and the quality and timing of medical care all affect the probability and severity of impairment, and the probability, extent, and pace of recovery. Many of these factors, along with motivation and family, community, and public support, influence how disabling any particular impairment will be in the labor market.

These observations about disability carry two important implications for the design of government policy. First, incentives matter. Poorly designed government policies can increase the proportion of the population experiencing work disabilities and can increase the economic costs of disability. Second, accurately assessing who is work disabled and to what degree is an illusory goal at best—one that has stumped federal bureaucrats, administrative law judges, and federal judges dealing with social security disability cases for the past three decades.

An enormous information problem is inherent in deciding who is disabled, what the effects of disability are, and how those effects can be minimized. Solutions require utilizing the market and the information and incentives it creates.[34] A drawback of the civil rights–reasonable accommodation approach is that it seeks to standardize market outcomes, thereby limiting the benefits to be derived from the market process.

Some Concluding Thoughts

Promoting the employment of the disabled requires expanding the number of disabled people willing and able to work and improving the information and understanding of employers concerning the skills, capabilities, and potential of people with disabilities. (Surely if parents and educators are still being awakened to the potential of disabled children, then employers are likely to need some awakening too.) Public policy should foster an environment in which disabled people have the incentive and the opportunity to gain market skills and in which employers have the incentive to develop creative new ways of incorporating the disabled into their work forces. Incentive-based approaches to reform hold real promise for promoting the well-being of the disabled at lower economic costs than mandated benefit programs.[35]

In thinking about the role of government in promoting the employment of the disabled, it is easy to overlook the government's wider responsibility to ensure a sound and growing economy. Economic growth may well be the single most potent weapon for creating new jobs and new opportunities for some of the nation's most disadvantaged citizens. Let us hope that the burdens imposed by the ADA on American businesses do not undermine the achievement of sustained growth.

2
Disability Accommodation and the Labor Market

Sherwin Rosen

Mandated workplace accommodation for disabled workers has become law with the enactment of the new civil rights bill for the disabled, the Americans with Disabilities Act.[1] Advances in medical care, increasing life expectancies, and technological changes that have improved the quality of life for many disabled people have at the same time increased their relative political strength and brought their plight squarely before the public eye. Yet public policy toward the disabled can take many forms, not all of which are equally worthy. The job accommodation mandate in the ADA raises important questions: How is it likely to affect the labor force activities of disabled people and the demands by firms for both disabled and nondisabled workers? Who will pay the costs?

In modern economies labor markets serve at least two important allocative functions. One is to distribute given labor resources among alternative uses. In this the market helps solve a matching problem, assigning people of different tastes, talents, and personal investments to varying occupations, industries, and firms. The other function is to determine the total labor resources used in the market economy—the total allocation of time between work and nonwork activities. For an individual this is the problem of labor supply: How much time should be devoted to working as opposed to other things? For a firm it is the problem of deciding how much labor to use in production along with other factors of production, and how to design an efficient work schedule for its employees. Appropriately enough, much previous research attention has been devoted to the connections between disability and labor supply. Job accommodation mandates, however, will affect both allocation problems; analyzing the incidence of what

I am indebted to Gary Becker and Carolyn Weaver for advice and criticism and to the National Science Foundation for research support. I alone am responsible for the views put forth here.

amounts to a mandatory employment subsidy for disabled workers requires a consideration of both aspects of labor markets.

Labor Force Participation of Disabled Persons

Economists have established a close connection between disability and labor supply. The explosion of disability payments and the rise in reported work-preventing disabilities contributed to the decline in male labor force participation in the 1970s.[2] The disability–labor force participation connection is sustained by the nature of the social security disability insurance (DI) program, where indemnification is contingent on a certified disability event that is itself defined in terms of work activity restrictions. The growth in DI payments has been caused in part by benefit increases and in part by inappropriate administrative decentralization. Indemnity eligibility decisions are made at the state level, but the program is financed at the federal level. Lack of tax constraints on states' spending decisions has had serious moral hazard consequences, including excessive growth in eligibility and unequal treatment of individuals living in different states.[3] Lack of alignment of financial interest and incentives between the two levels of administration has caused a host of incentive problems in the DI program that still remain.

The growing generosity of DI notwithstanding, the average person who receives disability payments is in poor health.[4] The increase in spending on DI over the past two decades undoubtedly has improved the quality of life for eligible disabled persons, but it hasn't done so in an efficient and equitable manner. Our disability system should promote incentives for disabled persons to make efficient work decisions. Both financial considerations and the availability of quality work opportunities are important to this choice. Obviously, financial incentives to seek employment rather than accept insurance payments will interact with the degree of accommodation to disabled persons offered by the job market. This interaction will determine our effective social policy, as the likely effects of job accommodation cannot be evaluated accurately without considering the insurance problem. Unfortunately there is a fundamental conflict between accommodation and insurance: one promotes work choice and the other discourages it.

Consider a few facts about the disabled and the labor market. Some of the most informative national data available on the incidence of disability come from self-assessment of work-limiting disabilities in census records, which is reviewed in a study by Robert F. Cotterman and John Raisian.[5] Based on these data, disabled persons accounted for approximately 13 percent of the total U.S. civilian popu-

19

lation older than fifteen years in both 1970 and 1980. (Advancing life expectancy among the elderly should have led to greater numbers of disabled elderly persons in 1980, but work-limiting disabilities are less likely to be reported for this group.) In 1980 almost 70 percent of adults with disabilities reported that their disability prevented all work activity. These people are out of the labor force, neither working nor seeking gainful employment. Of the 30 percent who did work, more than two-thirds spent at least thirty-five hours per week on the job. Thus, work among those who report disabilities appears to be an all-or-nothing affair: most don't work at all, but the minority who do work seem to hold full-time jobs in those weeks when they are employed.

Cotterman and Raisian calculate that during the decade of the 1970s the proportion of the disabled not working increased among virtually all age and sex groups (except for white females, where the proportions fell slightly against a background of rapidly rising labor force participation rates). The growth of social security disability payments has been directly linked to this phenomenon, as the increasing number of disabled persons not working was accompanied by a rapid increase in transfer payments to virtually every age and sex group.[6] Notwithstanding this increase in payments to the disabled, it is noteworthy that there was little change in the incidence of poverty among the disabled. Apparently income from transfer payments and insurance was substituted for income from wages and salaries as people either reduced their work activities or altered their family situations to become less dependent on others.

In summary, the background of the ADA is one of significant labor force withdrawal by disabled persons, accompanied by increased transfer payments and expanded benefit eligibility through the social insurance mechanism. Most disabled persons do not work. Whether eliminating discrimination against the disabled and mandating job accommodation will encourage labor force participation by disabled persons remains an open question. The effects will depend on the nature of the accommodations required by the legislation, the type of disabling conditions covered by it, and the work incentives implicit in the availability and comprehensiveness of social and private insurance.

Discrimination and Accommodation

The employment provisions of the ADA contain serious ambiguities that will have to be resolved by the courts over the years. One is the definition of "reasonable accommodation," which is vague and unspecified. Another is the meaning of "discrimination."

To economists, and to many other people, discrimination means either that identical people are treated differently or that unequal people are treated equally. The Civil Rights Act of 1964 is based on the view that differences among persons according to race, sex, and religious beliefs are socially inappropriate bases for discrimination, even though they might enter into employer, consumer, and worker preferences. Under the law, such markers must be ignored in determining equality of treatment in the labor market and elsewhere. Other and more directly observed productivity attributes can be used to make economic distinctions among workers. The willingness of most people to comply with a law that bars employment preferences on the basis of tastes or distastes for race, sex, or religious attributes probably lies in the fact that most people now consider these attributes irrelevant for determining labor market success. With actions on the basis of tastes made illegal, the Civil Rights Act has greatly elevated the importance of productivity and other signals in the job market.

The Civil Rights Act has attacked racial and sexual prejudices in their root sense of preconceived judgments and opinions based on biased, incorrect information. In many ways it has been successful in this endeavor. Perhaps the ADA will work in this beneficial way for disabled persons as well, assisting in overcoming the tendency toward oversolicitous concern for persons with certain conditions and insufficiently solicitous concern for persons with other conditions. In this way the law may assist in eliminating a type of "statistical discrimination" and allow information on the productivity traits of specific disabled workers to be communicated to employers and fellow workers in an unbiased manner.

Nonetheless, there is an obvious difference between the earlier law and the one being considered now: race and sex need have no relationship to productivity per se, but that is not true of most of the disabilities covered by the pending legislation. Indeed, the need for costly accommodation arises precisely because of differences between disabled and nondisabled workers. Fundamentally the ADA is not an antidiscrimination law. By forcing employers to pay for work site and other job accommodations that might allow workers with impairing conditions defined by the law to compete on equal terms, it would require firms to treat unequal people equally, thus discriminating in favor of the disabled. If not by imposing hiring preferences then by imposing greater expenditures to produce the same amount of work, it would be discriminatory.

It will be unfortunate if the ADA forces employers to provide forms of social insurance to disabled persons in the guise of accommodation. Social insurance is best provided and monitored directly by the state through the tax system. The definition of covered disabilities is bound to be fuzzy at its edges, leading to needless uncertainties for employers. Likewise the linkages between program costs and the direct taxes that voters can see will be lost, making politicians less accountable for expenditures they vote for. The lack of an explicit record will lead to needless problems of political control.

Curiously, the ADA applies antidiscrimination protection to certain health conditions that most afflicted persons would take pains to conceal from others. These conditions include alcoholism, drug addiction, mental illness, and some communicable diseases. There are obvious reasons why workers would wish to suppress some personal information that does not necessarily elicit prejudice but bears on the costs imposed on others. The behavioral aspects of some conditions might make accommodation very costly indeed.

It seems unlikely that the ADA will affect the propensity of workers to declare many types of covered disabilities. Indeed, the revelation of many of these conditions conflicts with the "privacy rights" that, though considered dubious by some, are discussed so frequently by policy makers. Any financial incentives sufficient to induce a person to reveal certain conditions probably would be so large as to fall outside the category of reasonable accommodation. The reasonable accommodation requirement will probably be most beneficial for people with nonbehavioral permanent impairments, such as loss of senses, motor and limb impairments, and mental retardation, and will do far less for people with emotional or behavioral impairments such as mental illness or various addictions.

Will Labor Force Participation Increase?

Out of some 22.5 million self-reported disabled persons in the 1980 census, approximately 15 million persons did not work and approximately 7.5 million worked.[7] We have no definitive evidence on the extent to which mandated accommodation will change these proportions, but an economic analysis of work choice helps illuminate the possibilities.

Economic theory and common sense indicate that nonwork-income prospects play a big role in the decision to work by both disabled and nondisabled persons. This is why economists have examined the connections between disability benefits and labor supply. Growth of benefits has been found to discourage work.

The slower growth of benefits in recent years will reduce incentives for disabled persons to leave the labor market. Furthermore, the remarkable inventions that improve the quality of life and increase

22

the life expectancy of many disabled people may enable some for whom it was previously impossible to enter the job market. Yet life extensions caused by medical and other advances expose most people to greater risk of disabling conditions at later ages, and this probably cuts the other way. Finally, there have been changes in the types of jobs available in the labor market that should influence the work decision. In particular, there has been tremendous growth in service industry jobs. It is not yet known to what extent these jobs are inherently more accommodating to disabled workers than are those in the contracting manufacturing sector, but it seems intuitively probable that they are so.

On the demand side, potential disabled workers usually have less education and fewer labor market skills than other workers. During the 1980s labor market opportunities for less skilled workers declined markedly.[8] If this trend continues, it does not bode well for the employment prospects of disabled workers. Raising the number of disabled persons employed from 7.5 million to 9 or 10 million appears to be the most that could be hoped for.

From the employer's perspective, the costs of accommodating disabled workers will be a barrier to increasing employment opportunities. Only one study is available, that by Berkeley Planning Associates, and it indicates relatively small costs of accommodation in most instances.[9] This inference can be extrapolated to the labor market at large only with great uncertainty, however, because the study's sampling basis is insecure. Many firms did not respond to the survey, and many did not keep sufficiently accurate accounting records to be able to respond. Equally important, the survey restricted its inquiry only to the direct capital expenses of accommodation and ignored such indirect expenses as extra supervisory time, sick time, and extra training expense, which would be important for many jobs. This is an important oversight, since it is by no means clear how such expenses will be treated in determining reasonable accommodation under the law. For these reasons, the BPA study must be regarded as a liberal lower bound to these costs.

Some Economics of Accommodation

In economics, questions of incidence refer to the ultimate burden of a tax or subsidy, as distinct from whom or what the tax or subsidy is nominally levied on. A sales tax on some good is nominally levied on producing firms, for instance, but is really paid by consumers in the form of higher prices and by cooperating resources in the form of lower wages and rents. Insofar as employment accommodation

23

amounts to a subsidy to disabled workers, questions of incidence naturally arise that economic analysis can address.

Let us consider the standard economic criteria of efficient work choices—a Pareto optimum allocation of labor—and, to simplify the analysis, ignore any taste aspects of discrimination toward disabled workers. We can ignore, too, the question of how much cost will be deemed reasonable by the courts and the degree to which firms and workers will comply with the law, since these are unknowable factors at this point.

Economics recognizes two criteria for efficient employment. One, applying to those who work, requires that the social product of an additional hour of work must equal its additional social cost. The value of additional work must not fall short of the additional cost of work, including the opportunity costs of the worker's time and of cooperating resources used to produce the product. The other criterion, applying to those who do not work, reflects the global condition that the total social product of one's labor must fall short of its total cost; otherwise some social surplus would be gained by working rather than by withdrawing from the labor market. If there are no fixed costs of employment or of labor force participation, this second criterion is assessed by comparing the product of the first hour of work with its costs. When fixed costs are present, however, the global test cannot be avoided because the worker's marketable product must cover both fixed and variable costs of working. Under normal conditions, wage payments in competitive labor markets serve to implement these criteria and align social and private interests.

These two economic behavioral conditions elucidate the observed work activities of disabled persons. First, the propensity for disabled persons not to work results from the combination of greater fixed and marginal costs of work time and a smaller marginal product. On the cost side, many disabled persons bear extra energy and exertion costs for each hour of work, and this is an important cause of the decision to leave the labor market. A smaller physical product of work time is a compounding factor when, for similar reasons, a smaller volume of work can be accomplished in a given period of time. It is unattractive to participate in the labor market when work is onerous and its reward small. For those with long-term disabilities these factors reduce the incentives to acquire job market skills and to invest in human capital generally. The expected returns on investments are smaller and the costs of investing are greater than for other people. This is the reason many disabled people do not work at all and tend to have lower than average skills.

Note that the costs of work are broadly conceived to include the subjective value of foregoing nonmarket uses of time. There is much evidence supporting the idea that nonmarket uses of time are a superior good whose value increases with one's standard of living. That idea might seem less applicable to disabled persons because they often must expend extra effort and energy in nonwork as well as work activities. Nonetheless, there is every reason to think that leisure is a superior good for disabled as well as nondisabled workers. Increasing disability benefits, then, would naturally lead to labor force withdrawals, as indeed was observed during the 1970s. Only if technological improvements and medical advances were to reduce disabled persons' psychic and monetary costs of working and increase their market productivity, should we expect to see increasing labor force participation. Perhaps these factors have been offset by greater insurance support in recent years.

Second, greater variable costs of work in both time and effort may be associated with greater fixed costs of labor force participation. Walter Oi has persuasively argued that the greater personal maintenance time dictated by many handicaps effectively reduces the total time available to accomplish other things. If handicaps increase the fixed costs of participation, including such aspects as the effort to prepare for work, the setup time at the work site, and the transport costs in time, effort, and money to get there, then participation is discouraged. But among those who do work, part-time work is a much less attractive option. Spending only a little time at work is not attractive, either financially or psychologically, when a person has to pay large fixed costs for the opportunity. Fixed costs discourage short hours of work and encourage longer hours. That the majority of disabled employed persons according to self-reported census data work thirty-five hours or more per week is best understood on these terms.

Costs and Benefits of Accommodation

Legislation mandating job accommodation will encourage positive work choices by the disabled because it will increase their personal productivity and reduce their personal costs of effort at the worksite. These positive work choices will be enhanced by public accommodation requirements that reduce the fixed costs of market participation as well as the costs of human capital investments. Certainly disabled persons as a group will be benefited by the ADA.

But that is not enough for good social policy. We also need to know how much these interventions will cost and who will pay the expense. A policy aimed at one group is unwise if it harms others more than it benefits the group at whom it is directed.

First, there is no inherent reason to expect that labor markets free of government intervention will fail to provide job accommodations in normal job situations. In an unrestricted, voluntary market situation, a worker would be both willing and able to pay the costs of accommodation if the efficiency conditions described above were satisfied. Then accommodation would occur precisely in those circumstances where it should occur—where its benefits exceed its costs. The worker would not be accommodated where accommodation is inefficient because costs would exceed benefits.

Determining precisely how accommodation costs are recompensed in private markets is irrelevant to the efficient work decision, though this can have practical consequences in terms of decentralized decision making. For example, firms could "charge" for the extra accommodation expense by paying a lower wage to disabled workers: this is equivalent to having the worker remit actual money payments for the extra use of time, training, and equipment involved. Or, as happens more often now, the worker could bear the training cost prior to employment and purchase the equipment on his or her own behalf. An accommodation law changes the form of payments but does not affect efficient work decisions unless accompanied by other restrictions.

The place where a reasonable accommodation law bites is in its so-called antidiscrimination aspects, which prohibit the compensatory mechanism from functioning through wage adjustments or other means. Then the burden of accommodation expense is shifted to others, and distortions in efficient resource allocation decisions result. Alternative arrangements for transferring resources would be preferable, unless the law changed tastes and sufficiently reduced discrimination against disabled workers.

How do these distortions come about and who bears their costs? It is easiest to think of a case in which disabled workers are perfect substitutes for some other workers, though not necessarily on equal terms on a worker-to-worker basis. The immediate impact of the law is to increase the demand for disabled workers and to decrease the demand for nondisabled substitute labor, thus lowering the wages of substitutes. The subsidy distorts the work margin because disabled workers do not face the full social costs of their work decisions—the costs of accommodation. Their private costs of working have been reduced, but the social costs are borne by others, so some will be employed in job situations where their productivity does not justify their cost. If nondisabled substitute workers are supplied inelastically to the market, the burden of accommodation costs is shifted entirely to them in competitive markets through a reduction in their average wage. Their average wage is reduced further if the increased supply of disabled workers entering the market is significant.

If the supply of substitute labor has some elasticity—some responsiveness to the wage—a further efficiency distortion is introduced. Some marginal nondisabled workers leave the market sector, where they were efficiently employed, for the nonmarket sector, where they are inefficiently employed. The switch allows them to escape the implicit tax that has been imposed on them, but it reduces true national income. Another way to escape the tax is to work slightly fewer hours during the year. Then the increased supply of disabled labor can be more or less offset by a reduction in the supply of substitute labor, depending on relative supply responses to wages in each group and on their relative proportions. Since disabled workers account for a relatively small proportion of the labor market (depending on the definition of disability), a total reduction in the supply of labor might be expected. Firms in the market sector are induced to use excessively capital-intensive production methods to economize on scarcer labor resources.

The incidence of accommodation subsidies will be shared unequally among different types of production and output markets. The greatest impact will be felt by industries that use more disabled workers and where the costs of accommodation are larger. The subsidy is not simply paid by firms in these industries. The higher costs will be passed on not only to other workers but also to consumers, in the form of higher product prices. Of course some industries specializing in the production of accommodation goods (wheelchair ramps, for example) will be aided by the ADA, encouraging socially excessive entry into these markets.

The effects of the law on different types of nondisabled workers depends on the degree of substitutability or complementarity between them and disabled workers. Part of the subsidy is shifted to those who are complementary with disabled workers because the demand for their services increases. This includes certain allied medical personnel and therapists, supervisory and training staff within firms, and those employed more generally in the accommodation goods sector. The costs of accommodation are shifted most to those able-bodied workers who are the best substitutes for disabled-worker labor, and the incidence of the costs is borne in proportion to the degree of substitution. Disabled workers tend to be less skilled than average, and so lower-paid, less-skilled workers probably will bear more of the costs. On the other hand the benefits are shared by complementary labor, which is probably no less skilled and may well be better paid than average.

Public Accommodations

Interventions in the area of public accommodations are said to have a rationale different from worker-specific job market interventions because they may involve certain externalities or scale economies. Potential market failure makes intervention more appropriate, in principle, if private bargaining and negotiation cannot achieve satisfactory results in these circumstances. If transfers are required, for example, it makes sense to provide lifts on buses only if a large enough share of buses are so equipped. This applies similarly for phone service for the hearing-impaired. Such cases of network externalities or scale economies inherently require public decision making through the political process. As the number of disabled persons who can potentially use these systems increases, their political voice also increases and they should (and do) play a more important role in these decisions.

One important point must be borne in mind, though. When production is subject to a scale economy, the prospective output must be large enough to be worth exploiting. Otherwise the cost of the project far exceeds its social returns, and public accommodation requirements are subject to potential abuses of the pork barrel variety, wherein the suppliers of public accommodation goods are the main beneficiaries of the law. This principle was largely missing from the debate surrounding public accommodations and the ADA. In many circumstances there is more than one way in which a public accommodation can be produced, and the nature of the accommodation chosen has to reflect the size of the population served. Mandating lifts in buses is costly business both in money outlays for society and in time losses for nondisabled passengers. If it makes economic sense at all it would be only in large cities, where many persons can make use of these systems. It is unwise in small cities, where few potential users reside. The distribution of disabled persons over cities and states in this country is not uniform because some climates and areas are much more amenable and inherently accommodating to people with disabling conditions. Uniform national standards are therefore inappropriate and will inappropriately impose much greater costs on some locales than on others. A better policy would be to allow local options in the choice of accommodation, such as special buses or taxi services where utilization is low.

Summary and Conclusion

There are three main points to be borne in mind when thinking about the employment provisions of the ADA. First, the legislation has been adopted against a backdrop of decreasing labor force participation and increasing insurance benefits and other transfers to disabled persons. These policies are inconsistent with accommodation policies aimed at encouraging labor force participation.

Second, the reasonable accommodation mandate in the private workplace will subsidize disabled workers and prevent the costs of accommodation from being appropriately priced and compensated. Under these circumstances disabled workers will not face the full social costs of their decisions, and their work choices will be socially inefficient, possibly including socially excessive participation and effort. Employers themselves will not pay for these accommodations; rather the costs will be shifted to other workers and to consumers. Workers who are closer substitutes for disabled workers will pay more of the costs in the form of lower wages or fringe benefits. Since disabled workers tend to be of lower-than-average skill, it is likely that the costs will be substantially shifted to lower-skilled nondisabled workers. Workers who are complementary to disabled workers, including those in the accommodation goods-producing industries and in some branches of the medical profession, will share in the subsidies. The reallocation of nondisabled labor resulting from these price distortions will lead to less efficient choices and lost national income.

Third, the reasonable accommodation mandate appears to be a politically easy way to shift a public burden off the budget at a time when the government budget is in deficit and there is widespread public distaste for greater taxation. This shift is ill-advised. Detaching this kind of legislation from the public budgetary process forfeits a crucial political control mechanism for limiting what actually are public expenditures. The costs are diffused, shifted, and hard to track down. They become invisible and can no longer serve to monitor the size and scope of the programs.

A stronger case perhaps can be made for government intervention in the area of public accommodation because some public services involve scale economies or network externalities. Even so, public accommodation standards must be sensitive to both local needs and relative numbers. Otherwise the distribution of the accommodation burden will be inefficient and inequitable, imposing increasing demands and distortions on local public finances. National policy must recognize that alternative production methods can produce equivalent outcomes and that local circumstances may favor different methods.

My analysis generally suggests that more direct subsidies to disabled workers for specific rehabilitation and training, financed by general tax revenues and open to public scrutiny wherever possible, are preferable to job accommodation as a matter of social policy. Vague and costly mandates on employers serve to spread the cost and benefits around the economy in unanticipated ways. They also conceal true social costs and give rise to labor market distortions that raise the total costs still further. Many of these issues are emotional and properly so, but that is not sufficient reason to shy away from fully understanding the probable consequences of the policies we adopt.

3
Disability and
a Workfare-Welfare Dilemma

Walter Y. Oi

Disability and poverty are important social issues that have historically defied satisfactory resolution. Part of the problem with respect
to disability is that the target population is an elusive and elastic one.
Statistical evidence shows that the incidence of work disability increased from 1960 to the early 1970s, dropped in the late 1970s, and
stayed level in the 1980s. The disability benefits awarded by the Social
Security and Veterans administrations, moreover, create work disincentives that have resulted in a significant decline in the male labor
force participation rate. More men are leaving the labor force, going
to the disability rolls, and staying there. In the late 1970s and early
1980s, Congress and the Social Security Administration took steps to
protect the fiscal integrity of the social security disability insurance
(DI) trust fund. It got harder to qualify for disability benefits, and
those on the rolls whose health conditions were subject to change
were required to undergo an eligibility review to confirm their right
to continued benefits. There was, in short, a recognition that disability is not an easily determinable state. Classification errors and
changes in conditions will occur. In addition, it was suspected that
some disability beneficiaries had the capacity to work and would do
so if they were properly motivated or pushed.

The disabled population has a dismal employment record. A
majority of all working age, disabled persons do not work. The
disabled who find gainful employment are more likely to hold part-
time jobs and earn hourly wages below those of their nondisabled
colleagues with similar characteristics.

Two hypotheses could explain this dismal employment record.
One is that medical impairments and functional–activity limitations
reduce productivity. People with disabilities want lighter work and
may choose to retire from the labor force. These preferences are, I
shall argue, a consequence of the fact that disability steals time.

Disabled persons need more sleep, more time for personal care and other chores, and more time for what Michael Grossman calls "investments in health capital"—visits to doctors and hospitals and rests to recover from illnesses.[1] The second hypothesis is that employers are uninformed about how productive the disabled can be and discriminate against them in the labor market. This hypothesis underlies the Americans with Disabilities Act (ADA).

Our earlier disability policies have mainly been directed to providing direct services or income maintenance. Social service and rehabilitation agencies offer disabled persons assistance in establishing routines for everyday living and train them for reentry into the labor market. Transfer payments, previously channeled through churches and private charities, have been shifted primarily to federal and state governments.

The ADA represents a major departure in our disability policies. Congress drafted, and the president signed, legislation that creates rights for the disabled, granting them equal employment opportunities, equal access to public places and services, and by implication more equal incomes. Employers will be required to assume the cost burden for providing reasonable accommodations that will enable qualified disabled individuals to compete for jobs. Failure to do so will be construed as discriminatory. The provisions of the act will evidently be enforced by a combination of regulations and private lawsuits.

Legislated rights can be confused (and are often equated with) entitlements. When this occurs, aggrieved individuals are more likely to appeal to the state and the courts than to try to negotiate a mutually advantageous arrangement between employer and employee. The rhetoric of the last decade has emphasized the importance of independence for disabled people. Disabled people, we have been told, should be encouraged to control their lives not only in living arrangements, but in employment as well.

The ADA undermines this goal of independence. Disabled people cannot have it both ways. They cannot achieve true independence by demanding equal job opportunities and equal access when the social costs of these rights are thrust upon the general public. The act will not produce the results intended by its designers; it will instead create a class of disabled persons increasingly dependent on the state for their continued well-being.

The Disabled—An Elusive and Elastic Target

Disability is an elusive state that cannot be precisely defined or measured. Functional and activity limitations surely matter, but there is no bright line separating the disabled from the nondisabled.[2] We can all describe the kinds of people who are disabled, but for policy purposes we need an operational definition. The ADA embraces the same definition as that in the Rehabilitation Act of 1973:

> The term disability means, with respect to an individual, a physical or mental impairment which substantially limits one or more of the major life activities of such individual; a record of such an impairment; or being regarded as having such an impairment.[3]

It differs from the statutory definition for the DI and supplemental security income (SSI) programs, which define disability as

> the inability to engage in any substantial gainful activity, SGA, by any medically determinable physical or mental impairment which can be expected to result in death or has lasted or is expected to last for a continuous period of not less than twelve months.[4]

The emphasis here is on work capacity, which is evaluated by a disability determination service administered by the states. Disputes arise and denials are appealed.

The DI program was introduced in 1956 to provide benefits for covered workers who were no longer able to work. SSI, a means-tested program, was established in 1974 to consolidate and extend income support to low-income disabled persons who could not qualify for DI. Each DI or SSI recipient is permitted to earn up to $500 a month without losing any benefits, but earnings in excess of this are effectively taxed.[5]

The social security disability programs thus create a work disincentive, encouraging working people with disabilities to drop out of the labor force and nonworking beneficiaries to remain out of the labor force. As benefit levels rise the number of disabled beneficiaries expands, and the male labor force participation rate declines. Between 1955 and 1985, for example, the percentage of forty-five- to fifty-four-year-old men not in the labor force rose from 2.5 percent to 8.2 percent; among fifty-five- to sixty-four-year-old men it climbed from 12.1 to 32.1 percent. Parsons and Leonard have argued persuasively that a significant part of the decline in male labor force participation rates is attributable to the increased generosity of disability transfer payments.[6] The statistical evidence supports the proposition that the target population of severely disabled persons who are unable to undertake any gainful activity is an elastic one.[7]

The Bureau of the Census relies on a combination of self-assessment and program participation to identify the target population.[8]

The data shown in the top panel of table 3–1, which are based on the Current Population Survey, reveal that the percentage of adults with a work disability increased from 7 percent in 1960 to 11 percent in 1973, then declined to 9.5 percent in 1984. The lower panels exhibit no trend; over the last eight years, 1981–1988, 8.75 percent of working-age adults reported having a work disability, and roughly half of

TABLE 3–1

PERCENTAGE OF WORKING-AGE ADULTS
WITH A WORK DISABILITY, 1962–1988

Year	Male	Female	Total
Haveman and Wolfe Estimates, Work Disability			
1962	9.5	4.8	7.0
1968	13.0	8.2	10.5
1973	12.8	9.3	11.0
1976	14.6	7.5	10.9
1980	11.9	9.6	10.7
1982	10.6	9.1	9.6
1984	10.5	8.6	9.5
Census Estimates, Work Disability			
1981	9.5	8.5	9.0
1982	9.3	8.5	8.9
1983	9.0	8.3	8.7
1984	9.2	8.1	8.6
1985	9.2	8.4	8.8
1986	9.4	8.2	8.8
1987	9.1	8.1	8.6
1988	8.7	8.4	8.6
Census Estimates, Severe Work Disability			
1981	4.8	4.7	4.7
1982	4.6	4.8	4.7
1983	4.6	4.7	4.6
1984	4.8	4.4	4.6
1985	4.7	4.4	4.5
1986	5.0	4.5	4.7
1987	4.8	4.6	4.7
1988	4.9	4.6	4.8

SOURCES: For Haveman and Wolfe estimates, see R. Haveman and B. Wolfe, "The Economic Well-Being of the Disabled, 1962–1984" (University of Wisconsin, Institute for Research on Poverty, April 1989, Mimeographed), table 1. For census estimates, see Robert L. Bennefield and John M. McNeil, "Labor Force Status and Other Characteristics of Persons with a Work Disability: 1981–1988" (Bureau of the Census, series P-23, no. 160, July 1989), table B.

them had a severe disability. Although the dispersion of the prevalence rates is quite small in the 1980s, the fluctuations over the longer time span are indicative of the difficulties of defining and measuring the population of disabled individuals.

According to G. DeJong, A. I. Batavia, and R. Griss, the total population of working age adults increased by 38.1 percent over the period 1958–1984, whereas the population of persons with a work disability grew by 157.9 percent, resulting in a marked upward trend in the ratio of disabled to nondisabled persons.[9] Although part of this trend is due to the negative correlation between mortality and morbidity rates,[10] the larger part is a result of the changing standards by which persons are classified as having a disability in the health interview surveys. In short, the criteria that are employed to identify a population of disabled persons are unavoidably imprecise, and as a consequence, classification errors are sure to emerge.

The Case for Government Intervention

The onset of disability, whether at birth or at a later age, is almost always associated with lower earnings and incomes. Much of this is attributable to less employment. According to Bureau of Census data, in 1987 only 41.9 percent of all disabled working-age men held jobs during the year, and only 17.1 percent were full-time year-round employees. The corresponding figures for nondisabled men were 92.1 and 63.5 percent, as shown in table 3–2. The mean earnings of disabled employees in 1987 were only 63.4 percent of that of nondisabled workers. If, however, the comparison is confined to year round workers, the earnings of the disabled rise to 80.7 percent of that of full-time, nondisabled men. A similar pattern was found in a Louis Harris survey conducted for the International Center for the Disabled.[11] In 1985, only 24 percent of working-age disabled adults in the ICD survey were holding down full-time jobs. Another 10 percent were working part-time, and 11 percent were unemployed but in the labor force. Two-thirds of the disabled who were not working said they wanted to work. Fully 34 percent of the disabled were over fifty-five years of age when they became disabled. These individuals are far more likely to retire and withdraw from the labor force. In fact, the earlier the age at the onset of disability, the higher the likelihood of being gainfully employed.

The loss in wage earnings may be partially offset by transfer payments or by the labor supply responses of other household members. The 1970 census data reveal that the wives of men with some work limitations had higher labor force participation rates. This is consistent with the added-worker hypothesis. The wives of se-

35

TABLE 3–2
Work Experience and Mean Earnings of Men by Work Disability Status, 1987

	Total	16–24	25–34	35–44	45–54	55–64
Disabled						
Total number	6,701	671	1,247	1,308	1,190	2,285
Percent						
Worked	41.8	51.9	54.3	46.6	43.6	28.3
Year-round	17.1	10.2	23.5	20.4	23.5	10.5
Did not work	58.2	48.1	45.7	53.4	56.4	71.7
Unemployed 4+ weeks	6.2	5.6	4.9	7.8	6.2	6.4
Nondisabled						
Total number	69,063	15,691	19,659	15,528	10,285	7,900
Percent						
Worked	92.1	79.7	96.8	98.1	97.5	85.8
Year-round	65.2	26.6	74.3	81.2	82.7	65.3
Did not work	7.9	20.3	3.2	1.9	2.5	14.2
Unemployed 4+ weeks	0.9	1.8	0.9	0.6	0.6	0.6
Mean earnings, all employed men						
Disabled	15,497	6,463	14,102	18,388	20,385	15,187
Nondisabled	24,095	7,581	22,362	31,082	33,775	28,899
Ratio	0.643	0.823	0.631	0.592	0.604	0.526
Mean earnings, year-round employed						
Disabled	24,200	14,985	22,249	27,524	26,618	22,601
Nondisabled	29,994		25,637	34,223	36,681	33,116
Ratio	0.807		0.868	0.804	0.726	0.683

Source: See source for census estimates, table 3–1.

verely disabled men, however, had lower labor force participation rates, resulting perhaps from the need to provide more personal care, but those wives who did work worked more hours per year than other working wives.[12] Of course, not all disabled persons have families. Paula Franklin found that the onset of a disabling condition increased the odds of a marital dissolution.[13] When these other sources of income are considered along with earnings, family incomes are well below the median. As a consequence, disabled persons are far more likely to fall below the poverty line.[14]

In addition to its impact on earnings and incomes, disability also affects the demand for goods and services. Individuals with disabilities understandably spend more on medical care and probably spend less on recreation and travel. Table 3-3 reflects this higher demand for medical services. The disabled make nearly three times as many physician visits a year and purchase more than four times as many prescriptions as individuals with no activity limitations. Even a minor activity limitation is associated with a more than twofold increase in expenditures for hospital care, physician visits, and prescriptions. The rapid rise in the real costs of medical care imposes a greater

TABLE 3-3

HEALTH CARE UTILIZATION BY DEGREE OF ACTIVITY LIMITATION OF PERSONS AGED 18-64, 1977

	Activity Limitation			
Item	None	Minor	Major	Severe
Physician visits				
Number per person	3.6	7.2	8.0	9.6
Relevant expenditures	1.0	1.9	2.2	2.7
Hospital care				
Discharges	11.3	20.1	35.6	66.5
Average length in days	5.2	8.1	10.1	14.3
% hospitalized once	12.8	19.3	24.6	30.1
% hospitalized 3+ times	0.4	1.2	2.0	3.9
Relevant expenditures	1.0	3.8	4.0	6.1
Prescribed medicines				
Number per person	3.7	9.9	13.1	17.3
Relevant expenditures	1.0	2.8	4.0	5.1

SOURCES: G. DeJong, A. I. Batavia, and R. L. Griss, "America's Neglected Health Minority: Working-Age Persons with Disabilities," *Milbank Quarterly* (1989), table 4. For hospital discharge data, see the 1977 national medical care expenditures survey. For percentage hospitalized, see 1979 health interview survey, discharges per 100 persons.

burden on the disabled, especially those who are ineligible for Medicare or are not covered by private health insurance. The loss of earning capacity and the higher demand for medical care form a strong case for government intervention. Government policies must, however, be carefully studied to determine whether they can succeed in promoting economic efficiency and improving the well-being of the disabled.

On Time and Functional Limitations

Medical impairments and functional limitations are ordinarily consulted in determining whether an individual is disabled. Walking, hearing, speaking, seeing, lifting, and thinking obviously influence the range of tasks that an individual can perform. An inability or limitation in any one of these dimensions is, however, only a symptom. It does not identify the causal link that leads to lower wages and labor force participation rates. For analytical purposes, it is useful to describe disability as a state in which an individual's stock of health capital, K, falls below some critical level, K_0.[15] Health capital can affect the individual in at least three ways. First, it affects preferences and hence demands for goods and services. If the onset of disability (meaning a destruction of some K) increases the disutility of work, it will lead to a decrease in the supply of labor to the market. Second, an individual's productivity or wage will be positively related to health. A wage reduction caused by a loss of some health capital will result in lower earnings or withdrawal from the labor force.[16] Third, K will surely affect the amount of time, T, that an individual can devote to work in the market, work at home, or leisure.

Each of us confronts the same potential time endowment, T^*, of 24 hours a day or 168 hours a week. But the time required for maintenance of the human agent, T_m, varies. People in poorer health (physical or mental) usually need more sleep, and those with functional limitations often require more time for bathing, dressing, and getting around. As a consequence, they have less discretionary time, $T = T^* - T_m$, which can be allocated to work or leisure. This time constraint gets even tighter when we move beyond the diurnal cycle. Table 3–3 reveals that individuals with more activity limitations spend more time in hospitals and doctors' offices, which together with days of illness ought to be counted as maintenance time. Disabled persons also have higher mortality rates, meaning fewer potential years in a lifetime. Less discretionary time reduces the well-being of the disabled, raises the preference for part-time work, and increases the readiness for retirement.

The characterization of disability in terms of medical impairments and functional–activity limitations encourages the adoption of policies that try to modify tasks and environments, or to train individuals to make better use of their remaining, smaller stocks of health capital. Our portfolio of policies must be expanded to acknowledge the wide diversity of disabling conditions not only in terms of impairments and limitations (such as hearing- and seeing-impaired, paraplegic, and developmentally disabled), but also with respect to the amount of *time* that an individual can control and allocate to work and other activities.[17]

Equal Employment Opportunities—Access to Jobs

Frank Bowe has proposed a theory of thirds, in which one-third of the disabled are "out of the labor force," half of the remainder are unemployed, and only one-third find gainful employment.[18] While other surveys report different percentages, the bottom line is the same: substantial numbers of disabled people do not work. Advocates for the disabled point to the ICD-Louis Harris poll, in which two-thirds of the disabled without jobs said that they wanted to work, as suggesting that ignorance and discrimination on the part of employers are responsible for this dismal employment record. They contend that the disabled can be just as productive as the nondisabled. Evidence supporting this contention almost always takes the form of illustrative case studies. No statistical study with which I am familiar compares the average productivity of a random sample of disabled persons, in and out of work, to that for a control group of able-bodied individuals.[19]

Some disabled persons are allegedly denied an equal opportunity to compete for jobs because they require costly workplace accommodations. Title 1 of the Americans with Disabilities Act prohibits employment discrimination, which is broadly defined to include, among other things,

> not making reasonable accommodations to known physical or mental limitations of an otherwise qualified individual . . . unless such covered entity can demonstrate that the accommodation would impose an undue hardship on the operation of the business.[20]

Discrimination also includes denying an employment opportunity because the individual needs an accommodation. The nondiscrimination clause in the ADA differs from that in the Rehabilitation Act of 1973 in at least two important respects. First, it covers private employers (with fifteen or more employees) who do not have federal contracts. Second, and more important, private lawsuits apparently will be permitted to enforce violations of the ADA; violations under section 502 of the Rehabilitation Act of 1973 cannot be enforced by private lawsuits filed by aggrieved handicapped workers.[21]

The conventional wisdom is that disabled persons face discrimination both in finding a job and in the wages offered to them. The ADA follows the Civil Rights Act of 1964 and tries to remedy the situation by mandating what I shall call "job rights." A qualified disabled individual has the right to insist upon reasonable accommodations in order to compete for a job or a promotion.

How valid are claims that most nonemployed disabled individuals want to work, and that they can be just as productive as the nondisabled if they are only given a chance and provided with reasonable accommodations? Time is clearly a serious limitation, as evidenced by the results of a 1975 survey of 889 vocational rehabilitation clients who were unsuccessful in finding work.[22] Some 62 percent said they wanted "light work," 47 percent asked for reduced work schedules, and 40 percent expressed a preference for "flex time." These findings confirm the thesis that disability steals time. The time budgets of some disabled persons prevent them from meeting the rigid, full-time work schedules demanded by some firms. Does a refusal to modify a work schedule to fit the smaller and less certain time budget constitute a violation of the reasonable accommodation clause? Rigid schedules and team production are more important for efficiency in some employments than in others. An unwillingness or inability to conform to such schedules ought, in my opinion, to disqualify an applicant for such a position.

Under ADA, the employer is responsible for accommodation costs. But who will actually bear those costs? Consider an example in which Blue workers earn lower wages in a competitive labor market because they need special lights to do the work. It is economically efficient to place the cost burden on the Blues through lower wages because scarce resources are needed to supply the special lights. Without lights, the Blues are less productive than their Green colleagues. If a law were passed whereby the employer had to pay the same wages to Blues and Greens as well as provide lights and the employer faced a nondiscrimination clause, the costs of the resulting inefficient allocation of resources would be borne by consumers and the Greens.

The supporters of the ADA insist that accommodation costs little. They sometimes point to the 1982 study by F. C. Collignon and others, which found that in a sample of firms subject to the requirements of the rehabilitation act, the disabled worker required no accommodations in 51 percent of the cases. In another 30 percent, the accommodation cost less than $500.[23] However, Collignon cautioned that these findings could not be extrapolated to estimate the costs of accommodating an expanded population of disabled employees. There are significant cost differences that may be related to the type and severity of the disabling condition and the nature of the job. Accommodation costs were higher for blind and paraplegic workers.[24]

In evaluating the accommodations clause of the rehabilitation act, Collignon raised two questions: "First, [is] the law . . . reasonable and effective in achieving its intent of facilitating the employment and integration of disabled persons into society, and second, who should appropriately bear the cost burden of the law?"[25] These questions are no less important today in evaluating the wider range of workplace accommodations mandated under the ADA.

Equal Access to Services and Places

Many individuals are denied access to some public or quasi-public services and places. Access can be improved by removing architectural and transportation barriers or by supplying supplementary aids and services. Under title 3 of the ADA, equal access will be guaranteed not only to public places but also to certain private places, such as restaurants, doctors' offices, pharmacies, grocery stores, museums, and homeless shelters, which under this act are presumed to be public places. Under this title, a firm or agency is engaging in a discriminatory act if it provides an opportunity that is

> different or separate from that provided to other individuals, unless such action is necessary to provide the individual or class of individuals with a good, service, facility, privilege, advantage or opportunity that is as effective as that provided to others.[26]

These rights to equal access impose mandatory standards on certain firms, similar to those under the fair housing act. They also prescribe the way in which some services (notably transit services) have to be produced.

Under the ADA, elevator lifts will have to be installed on all newly purchased transit buses. The facts here are nearly identical to the trans-bus case of 1979. In that case, the Congressional Budget Office reported that the trans-bus would be utilized by only 7 percent of all severely disabled persons at a cost of $38 a trip. It analyzed two alternative plans. Transit services would be supplied by a mix of kneel-down buses, dial-a-vans, and taxis, which would serve 26 and 30 percent of all severely disabled individuals at $7.52 and $7.33 per trip. These alternatives provided services that were nearly identical to the fixed route services of the trans-bus, but because they were "separate and equal," they were unacceptable to the advocates of equal access. Costs are irrelevant, and substitutes are unacceptable to the hard-line advocates.

In addition to the direct incremental costs (for the lifts and the wages of drivers and maintenance crews), elevator lifts will impose implicit costs on other bus riders.[27] Given their high capital costs, new buses fitted with elevator lifts will reduce the flow of federal funds available to subsidize urban transit systems. As a consequence, regional transit authorities will be forced to cut back on their fixed route services. The net result could be fewer transit services for all users, including the wheelchair riders.

It is argued that if places and services are publicly funded, justice calls for them to be accessible to all. In principle I am in agreement with this position, but common sense should be used to define a standard of "reasonably equal access." A newly constructed two-story motel has to be accessible to the disabled, but reasonably equal access can be achieved by insisting that only the first-floor units be made wheelchair accessible. Elevator lifts on all new buses are not a cost-effective means of improving the mobility of the disabled. The funds could be better spent on other programs. Social costs and benefits cannot be ignored, and alternatives should be carefully considered.

Equal Access to Adequate Lives

A majority of disabled men supply no labor to the market. They get their incomes from transfer payments or the earnings of other household members. A colorable case can be made to support the proposition that individuals have a right to disability benefits under worker compensation or DI, or from the Veterans Administration. They participated in an insurance program and happened to be the losers who became disabled, thereby qualifying them for insurance benefits. Others who get SSI benefits might claim that they participated in the larger social insurance program in which society assumed the responsibility for providing them with an adequate standard of living if they were unable to work. None of these, however, is an actuarially sound insurance program, and all invite a moral hazard risk. The existence of a social insurance program affects the likelihood that an individual will apply for benefits under the program. Policy makers are fully aware of these moral hazards and their associated work disincentives. It is an age-old dilemma in social welfare policy for which there is no easy solution. A desire to provide the disabled with adequate incomes is somehow balanced against a harsh budget constraint.

Rights or Entitlements—Independence or Dependence?

The DI and SSI programs take a rights approach, whereby a disabled person who meets the statutory definition has a right to benefits and an adequate income. Rights can easily be confused with entitlements, however; the boundary is fuzzy. When DI costs exploded in the 1970s, for example, Congress and the Social Security Administration were forced to take steps to limit program size by making it harder to qualify for benefits and forcing incumbents to reestablish their eligibility. Many on the rolls understandably believed they were entitled to benefits and so, having been granted them in the past, should not have been forced to undergo the cost and humiliation of disability review.[28]

Similarly, the right to equal employment opportunities under title 1 could be easily confused with an entitlement. We are not provided with operational definitions that describe who is a qualified disabled individual or what constitutes a reasonable workplace accommodation. Imagine a situation in which worker A applies for a position and tells the job interviewer that in order to perform the job in a satisfactory fashion he needs special equipment and the work schedule will have to be adjusted to fit in with his car pool. Twelve applicants are interviewed, and the job is offered to worker D. Worker A sues and claims that the firm discriminated against him because it would have incurred higher costs to provide the necessary accommodations. In its defense, the firm's lawyer argues that A and D have identical functional limitations (the accommodations were unreasonable because D didn't need them), and besides, D had stronger references (A was less qualified than D). Here, the defendant should prevail on the ground that an employer has the right to enter into a voluntary employment relation.

There are surely valid cases in which discrimination is responsible for a refusal to hire a qualified disabled person. Damages for cases involving a refusal to hire are sometimes difficult to determine because the reason for the refusal is not always clear. Disabled persons typically spend more time searching for a good job match and thus accumulate more refusals. We have guidelines for what constitutes a violation of equal employment opportunities, but we have no bright line to decide when an accommodation is unreasonable. The hard case is one in which a plaintiff sues on the grounds that an employer's failure to provide and pay for suitable equipment was responsible for placing him in a low-wage job and preventing his promotion. Courts will have to decide if the plaintiff is disabled and truly qualified for the higher paying position, and what constitutes a reasonable and sufficient accommodation. Damages in this latter case could be substantial if the employer has to pay back wages plus compensatory and punitive damages.[29] I predict that the passage of the ADA will lead to the creation and rapid growth of two new industries—one consisting of consultants who will interpret the legislation and regulations for wary employers, and a second made up of cadres of lawyers rushing to the bar to protect the rights of aggrieved qualified disabled individuals on the one hand, and to guard the deep pockets of the capitalists on the other hand.

A better approach, it seems to me, would be to modify our largest income support program for the disabled, DI, so as to facilitate return to work among those willing and able to do so. It is well known that some DI recipients are afraid to return to work in spite of improving health because they would have to give up Medicare. These people come from the lower income and education groups and are in poorer health. B. V. Bye and G. F. Riley report that some 27.8 percent of persons joining the DI rolls between 1972 and 1981 had no medical insurance in the six months before entry into the DI program.[30] In addition, they estimate that eliminating the two-year waiting period for Medicare benefits would raise Medicare costs by 45 percent. The risks of incurring unanticipated medical expenses pose a major deterrent for those considering a return to work. To promote work among current and prospective DI beneficiaries, some thought should be given to unbundling cash and Medicare benefits.[31] Under a scheme that seems feasible to me, an allowed applicant could be awarded either cash and Medicare benefits, or Medicare benefits only. Those who receive Medicare only could return to work and ask their employer for a pay package that excludes the usual medical health fringe benefit. The employer is thus provided with an incentive to hire the disabled worker who now entails a lower total labor cost. This implicit wage subsidy could be a powerful incentive to hire the disabled.[32]

Concluding Thoughts

"Towards Independence" is the title of the first annual report of the National Council of the Handicapped.[33] It is a long laundry list of policy recommendations that might improve the well-being of the disabled, some of which were incorporated into the Americans with Disabilities Act. The strategy is to expand the set of rights available to the disabled, which is at odds with our earlier policy approach, which emphasized the delivery of direct services. Under the ADA, disabled persons are guaranteed the rights to equal employment opportunities, equal access, and if unable to work at all, rights to an income stream that can sustain an adequate and equitable standard of living.

These rights will be enforced by a combination of civil lawsuits and regulations. Broadly defined rights may be erroneously equated with entitlements. A poorly defined and fragmented target population of disabled individuals who are told that they are entitled to jobs, access, and disability benefits could emerge as a cohesive, articulate, and vocal minority who will become increasingly dependent upon the federal government for their continued well-being. In order to enforce compliance with their legislated rights, disabled people will have to turn to lawyers and regulators. More and more resources will be allocated to establishing and enforcing rights and rules. The ultimate equilibrium, I suspect, will be the same as that attained in the case of "Jarndyce and Jarndyce," so eloquently described by Charles Dickens in *Bleak House* in 1851. The little boy who was promised a rocking horse when the estate was settled grew up to ride his own horse and went on to another world without ever seeing a farthing of the estate. The case was finally closed when all of the proceeds from the estate had been safely moved into the pockets of the barristers. The consultants and lawyers will do well, along with a few successful litigants.

Disabled people cannot have it both ways. They cannot simultaneously demand independence and insist upon guaranteed jobs that may, at times, entail accommodation costs exceeding the value of their product, public services, and adequate incomes. The ADA will result in an inflated population of disabled persons whose welfare will become increasingly dependent upon an ever growing federal bureaucracy. Charles Murray may have another opportunity to test his model.[34]

4
Disability Policy and the Return to Work

Jonathan S. Leonard

Many of the problems faced in disability policy arise from a fundamental oversimplification: disability is considered an immutable and obvious condition. Either one is disabled as a result of medical conditions and unable to work, or else one is not. To see that this premise is wrong as well as too simple, one need consider only the growth of the largest disability program, the social security disability insurance (DI) program. Section 223(d) of the Social Security Act defines "disability" as the inability to engage in any substantial gainful activity by reason of a physical or mental impairment that is expected to end in death or to last at least twelve months. Since the establishment of DI in 1954, increases in the beneficiary rate have been closely paralleled by decreases in the participation rate of prime-age men in the labor force. As more prime-age men have become beneficiaries, fewer men have worked or looked for work.[1] If the simple assumptions about disability were true, one would expect no such relation. If the disabled could not work, they would be out of the labor force irrespective of the DI program. And if only the disabled were admitted to the benefit rolls, the enlargement of the rolls would simply transfer income to these people rather than reduce the number working. Policy has had to confront a more ambiguous and complicated reality. Disability is not purely medically determined. Nor is disability determination a simple or an error-free process. There is great scope for economic incentives to affect behavior even in this area usually considered black or white. Likewise, there is great scope for errors in disability determinations.[2]

DI and the Return to Work

DI beneficiaries rarely return to work. Once initial eligibility is established, the program resembles an early retirement program. The Social Security Administration's chief actuary reports that in 1987

46

fewer than 8,000 disabled beneficiaries terminated benefits because they successfully completed a trial work period.[3] This represents less than one-half of 1 percent of all DI beneficiaries. Some 20,400 people were in their second and third years of Medicare eligibility while engaged in substantial gainful activity, a substantial increase over the previous year but still a small fraction of disabled workers.[4] The low rehabilitation rate may indicate that those disabled severely enough to enter the rolls are likely to stay disabled, but there is concern that the rate is a response to the strong incentives to stay on the rolls after some ability to work is recovered.

As in the process of screening for entry into the DI program, screening for exit is beset by problems inherent in a rigid conception of disability as a total and long-lasting condition that is medically determined. To see this, suppose that the world really operates in this simple black-and-white fashion: people are either disabled and incapable of work or else they are not disabled and are capable of work. In addition one could assume that the two states are easily determined and verified and change infrequently. Under these conditions a disability program can be relatively simple to run. One simply mails checks to the disabled and leaves the non-disabled alone except for tax collections. This system is fair and compassionate and easily shown to be both. By definition there can be no inefficiencies created by giving benefits to the disabled because their labor supply is inelastic and fixed at zero. These assumptions seem to underlie the DI program as established in 1954.

Most of the problems for policy arise because this black-and-white policy is forced to deal with a gray world. It does so poorly.

Historically the DI beneficiary rolls have expanded and labor force participation rates have declined among affected groups as benefits have become more generous and as eligibility standards have been loosened.[5] A number of avenues have been proposed and some implemented to limit program growth. These include reducing the generosity of benefit levels, tightening eligibility standards, increasing the accuracy of disability determinations, rechecking beneficiaries for continuing disability, and removing state influence over disability determinations.

How might DI beneficiaries be encouraged to return to work? One way is by making beneficiary status less attractive—for example, by reducing benefits. Only once has Congress taken this tack—in 1980—and then only by limiting the maximum family benefit. Another way is by reviewing periodically the continuing eligibility of people on the rolls and removing those who have regained the ability to work. The lasting residue of the most recent attempt to do this, in the late 1970s and early 1980s, is the reluctance of the government to burn its hand again. Before another attempt to purge the rolls, there is a need for increased public awareness of the difficulties of making disability determinations and of the incentives the current system gives to "gray" people—those not obviously disabled or not disabled—to apply for benefits, to the states as agents of SSA to grant them, and to beneficiaries to remain on the rolls.

If politically being a beneficiary cannot be made less attractive, and the public is unwilling to remove people from the rolls who have regained the ability to work, the only remaining alternative is to make work more attractive. The system could subsidize beneficiaries who return to work or employers who hire them, for example, or employers could be required to eliminate barriers to the employment of the disabled. These options are considered in turn.

Extended Medicare Eligibility

One disincentive to leave the rolls is embodied in current regulations. Social security provides not only cash benefits through DI but also health insurance benefits through Medicare. A beneficiary considering a return to work must contemplate not only the loss of cash benefits but also the eventual loss of Medicare coverage. For many people with potentially disabling conditions, the latter is more valuable and more difficult to replace. Medicare expenditures for disabled beneficiaries currently average about $4,000 per year.[6] Private insurance usually limits coverage of preexisting conditions or is simply not available to those with certain conditions. When insurance can be obtained, it sometimes costs the disabled so much that there is little financial reward to working.

Even though the individual might not benefit financially by returning to work, the government might, particularly if it lacks the will to remove people who can work from the rolls. One could suppose the government extended long-term Medicare eligibility to all those who had ever established eligibility for DI, irrespective of whether they were working or disabled. Under this regime the implicit tax on leaving the rolls would be just the lost cash benefits. An individual who succeeded in finding a job that paid in excess of these benefits would no longer have to worry about losing Medicare coverage upon return to work. This would reduce barriers for DI beneficiaries who want to work. This might be the rarest of all policies: one that reduces federal outlays by being more generous. Of course this conclusion would be off the mark if the offer of long-term Medicare coverage, which would increase the attractiveness of becoming a DI beneficiary, simultaneously swelled the beneficiary rolls.

Unfortunately one cannot turn to historical evidence to discern the likely response of beneficiaries and potential applicants to Medicare extensions. Medicare coverage was liberalized twice during the 1980s, but eligibility review policy whipsawed through a complete revolution: from virtually no reviews to pervasive and stringent reviews and back to practically no reviews. These flip-flops in policy dominate the statistics on the numbers of people leaving the rolls and cloud any attempt to discern the impact of changes in Medicare coverage on either applicant or return-to-work rates.

Today there is little prospect of being removed from the rolls because a person is found to be no longer disabled. If a beneficiary who chooses to return to work is covered by Medicare for four years. Hardly any DI beneficiaries return to work. Extending the current four years of Medicare coverage to long-term coverage might have a significant effect, but this cannot be inferred on the basis of the available historical evidence.

The more serious objection to this proposal is its codification of an enlarged conception of disability—and only at one end of the process. This proposal clearly envisions working people receiving Medicare benefits established for the disabled. This simply does not fit under a black-and-white model of disability. It acknowledges the existence of people who were once disabled enough to establish eligibility, who are still not so clearly able to work that the government is ready to remove them from the rolls, but who nevertheless do work. Partial benefits, in the form of Medicare coverage, are paid to these gray people who choose to return to work. The inconsistency is that these same gray people could not gain admission to the program; their work itself would disqualify them. While this may seem unfair, it is not unlike the way the current system is being operated.

A model for paying benefits on the basis of partial disability already exists in the veteran's disability program. Benefits are paid for service-connected disabilities in relation to the severity of impairment and irrespective of other earnings or income. One advantage of this more flexible definition of disability is the reduced pressure on gatekeepers (doctors, administrators, administrative law judges) to fit square pegs into round holes. It gives administrators more room to match benefits to the degree of disability. Gross inequities are less likely because people who differ only slightly in the degree of disability are less likely to be labeled at the extremes of either complete disability or complete ability to work. Appeals of disability determinations are less common under VA than DI regulations.

On the cost side a much larger proportion of the population is eligible for some benefits. Under likely conditions the DI program would become more costly—not a characteristic associated with proposals likely to survive in the current budgetary climate.

Section 1619 of the Social Security Act also provides a more flexible model. This enables disabled recipients of supplemental security income (SSI) to continue receiving Medicaid coverage even while earning more than the amount that would demonstrate "substantial gainful activity." By September 1988, 33,155 were enrolled in the section 1619 program—fewer than 2 percent of the disabled SSI recipients.[7] Evaluations suggest that section 1619, and the income-tested phase-out of benefits under SSI, have increased the number of recipients who return to work, although not in great numbers.[8]

An alternative approach, adopted by Congress in 1989, is to allow the disabled to buy Medicare coverage. This approach recognizes the potentially high cost or restrictive conditions of private insurance for severely disabled people attempting to return to work. The problem with this approach is that, with uniform prices, those expecting the greater medical costs will find Medicare the most attractive. But an actuarially fair price for Medicare coverage for such a population would likely be so expensive that few would be attracted to work by its provision. Any lower price would require large and open-ended government expenditures. Under the provision adopted by Congress, the price of Medicare for disabled workers equals the price charged to the uninsured elderly. It remains to be seen how effective this costly subsidy will be in helping the disabled return to work.

Civil Rights for the Disabled

An alternative approach for encouraging work, or a return to work, is enforcing civil rights for the disabled. Where discrimination has been seen as a root cause of employment problems, the government has taken the direct approach of outlawing discrimination. This policy has had some success in improving employment opportunities for minorities, women, and older workers. The Americans with Disabilities Act (ADA) extends similar protections against employment discrimination to those with disabilities.

The ADA is fundamentally distinct from other antidiscrimination laws, however, in its adoption of the principles of reasonable accommodation and undue hardship. These point to an important sense in which the ADA is not solely concerned with discrimination as the term is understood by economists and typically by the courts. Under the Civil Rights Act of 1964, discrimination is taken to be the unequal treatment of equally productive workers. But the explicit consideration of the cost of accommodation in the ADA emphasizes the fact that in some cases even employers who do not discriminate against the disabled may prefer not to employ them because of additional costs.

The ADA states the factors to be considered in determining whether accommodating a disabled person would be an undue hardship for an employer; these include the size of the business and the size of its budget. This is a far cry from cost-benefit analysis. A large enough company, or one with large revenues or profits, would presumably be expected to make quite expensive accommodations even if they would be of only marginal benefit to a few disabled people. By the same standard an employer with a small budget might be able to avoid relatively minor accommodations that would be very beneficial. The absence of cost-benefit language may reflect a desire to adopt the supposedly tried and proven language of the Rehabilitation Act of 1973. Court interpretations of the rehabilitation act, however, have been diverse enough to raise concerns. While some rehabilitation act cases have found any expensive accommodation to be unreasonable, thereby allowing employers to sidestep the act easily, the case of *Nelson v. Thornberg* adopts the deep-pockets principle: if the cost is small relative to the employer's total budget, the accommodation is probably not unreasonable.[9]

There is one major limit on this principle. The ADA appears to allow employers to provide insurance plans that classify the disabled into more expensive risk pools. Unless barred by state law, employers appear able to pass onto the disabled the extra insurance costs of their employment.

The ADA is also tempered by the absence of punitive or compensatory damages of the sort available under section 1981 to victims of racial discrimination. The remedies available follow the less onerous standard of the Civil Rights Act of 1964.[10]

These limited remedies represent part of a political trade-off for a broadened definition of disability that includes those perceived to be disabled. This is a fuzzy and potentially troublesome conception of disability. If a university decides not to hire someone because he is believed to be not smart enough, is that sufficient grounds for a claim of discrimination under the ADA and for a request that the university reasonably accommodate this real or perceived lack of mental acuity?

Affirmative Action for the Disabled

As another arm of a fragmented disability policy, affirmative action to encourage employment of the disabled has been proposed. The disabled, like other victims of discrimination borne of ignorance, are

51

assumed to be better able to gain employment with some gentle persuasion on the part of the government. If ignorance and discrimination borne of ignorance concerning the true work potential of the disabled were indeed a major source of the disabled's poor employment prospects, then affirmative action might help permanently improve these prospects by dispelling ignorance.

Although it is little noted, the federal government already has an affirmative action policy for the disabled, at least on paper. Section 793 of the Rehabilitation Act of 1973 requires contractors of the federal government to "take affirmative action to employ and advance in employment qualified handicapped individuals." The Office of Federal Contract Compliance Programs within the Department of Labor is responsible for enforcement. There exist no evaluation of the effectiveness of this program, no record of sustained vigorous enforcement, and nothing to suggest that the disabled seeking private sector employment have materially benefited from this provision. Simply put, the disabled are not and have not ever been a priority of the OFCCP.

Affirmative action is viewed with either hope or trepidation as a powerful federal policy capable of forcing great changes in the employment practices of the multitude of private employers throughout the country. The truth appears to be both more subtle and less imposing. The symbolic content of affirmative action policies has far outweighed their practical direct effect on employment. Even when affirmative action was at its peak during the 1970s, the actual employment changes engendered by affirmative action were small; they were practically nonexistent during the 1980s. [11] The program has enjoyed neither the clear sense of mission, nor the favorable consensus of public opinion, nor the funding necessary to implement its ambiguous goals fully. The existence of the program, if not its modest effects, is testimony to the intention of the federal government to accommodate the demands of minorities and women and to reduce discrimination by federal contractors.

In a curious but sad fashion, affirmative action for the disabled might be expected to meet with greater success than affirmative action for minorities and women. Racial and sex discriminations are exclusive risks. Members of one race or sex are by definition not direct victims of discrimination against other groups. In contrast disability can strike anyone. Disability and the consequent threat of discrimination against the disabled are inclusive risks. Everyone faces some risk then of becoming a victim of discrimination against the disabled. This fear, and the related vague guilt of those who survive whole, may help reduce opposition to affirmative action for the disabled. In addition the disabled are usually thought of as a small group who could be accommodated without greatly reducing the prospects of current employees. Their demands are not typically viewed as another manifestation of group power politics.

Many large employers demonstrate admirable records of employing the disabled, sometimes because it is good business, sometimes because it is good public relations. Some of the forces at work behind affirmative action policies for the disabled can be illustrated by comparing the treatment received by two identically disabled workers: the first, with a disability, applies for a job and the second becomes disabled after employment. The second is likely to enjoy greater success in employment, even if both would be equally productive on the job.

Many companies, for example, implicitly insure their current employees against misfortunes beyond their control by making accommodations for employees who become disabled. This form of implicit insurance is widely valued by employees. In addition, if the employer is obligated to pay increased disability benefits or health insurance premiums in any case, the net costs are reduced to the extent that the beneficiary remains at work.

In contrast, a disabled job applicant tends to be viewed as an outsider. Failure to offer that person the same implicit insurance is less likely to be observed or valued by current employees. And the employment of such a disabled applicant may raise rather than lower the employer's net costs if it leads to an increase in health or disability insurance premiums.[12]

While affirmative action and antidiscrimination policies seek to enhance the opportunities of disabled people, an unintended consequence may be to reduce implicit insurance for employees who become disabled. Title 7 of the Civil Rights Act of 1964 forced many employers to abandon idiosyncratic acts of charity and paternalism for more consistent and explicit policies based on business necessity that were more readily legally defensible. Under the ADA, employers will also eventually be forced to consider the consistency of their acts. The law may not require charitable acts, but it does require even-handedness. Any claim that it is too costly, too difficult, or unreasonable to accommodate a job applicant with a disability would be undercut by continuing to employ someone who develops the same disability once on the job. Corporate paternalism and charity have been known to wither rapidly under such conditions.

One major employer has a commendable record for employing the disabled: the federal government is probably the largest and most welcoming employer of the disabled. Something can be learned of employment barriers by considering not only where the disabled fail but also where they succeed in finding employment. A number of factors may account for the greater success the disabled achieve in finding federal rather than private employment. Some might credit a noble government that does not discriminate. Others might note that the government does not face the same competitive pressure to reduce costs. Significantly, the law affords greater protection to the disabled in government than in private employment. Section 791 of the rehabilitation act requires federal agencies to develop affirmative action plans for the handicapped. Section 2014 of the Veterans Readjustment Act of 1974 states that the policy of the United States is to "promote the maximum of employment and job advancement within the federal government for qualified disabled veterans."[13] This serves as a form of implicit insurance for those disabled serving the country, particularly in times of war. Finally the politics of compassion play a more prominent and direct role in government than in private decisions.

Conclusion

Public policy should aim to prevent disability where possible, to encourage the rehabilitation and return to work of those capable, and to provide for the welfare of those incapable of work. The history of past attempts to provide for the disabled through the social security disability programs demonstrates that as the generosity of public programs increases, so does the number of people who drop out of the labor force and cease looking for work. The proportion of disabled people is not purely medically determined: it responds to social and economic forces. A number of policies have demonstrated some success in encouraging the disabled to return to work. Others remain to be tried. To maintain incentives to work, benefit formulas can be (and sometimes have been) adjusted so that after-tax income is higher for people who work than for those who do not work. Extending Medicare coverage for the disabled who return to work is one attempt to increase work incentives, so far with modest impact. As the disabled population becomes younger and rehabilitation technology improves, more careful initial screening and subsequent recertifying of disabled beneficiaries should diminish the use of DI as an early retirement program, encourage the return to work, and strengthen public support for the DI program. This, in turn, should ensure that all who are worthy become beneficiaries, and that all beneficiaries are worthy. Screening errors cannot be eliminated, but they can be reduced to improve the fairness and consistency of the program.

A legal bar to employment discrimination against the disabled could potentially remove inefficient barriers to work. Policies that remove pure discrimination against the disabled are desirable on economic as well as moral grounds. The actual impact of such a law depends in part on how courts draw the distinction between the disabled who earn less because they are less productive and those who earn less because of discrimination, despite being equally productive.

Disability in Our Society

A Commentary by Evan J. Kemp, Jr.

The chapters by Carolyn Weaver, Sherwin Rosen, Walter Oi, and Jonathan Leonard offer some interesting arguments concerning the problems and prospects of the Americans with Disabilities Act (ADA). Although I disagree with a number of their conclusions, it is nonetheless exciting to see these scholars take this legislation so seriously. Too often legislation addressing the problems of disabled people has not been taken seriously, either by members of Congress, by scholars, or by others. Disability policy has been seen as charity for disabled people, not warranting analysis.

It is very difficult to define a disabled person. There are forty-three federal definitions and literally thousands of state, county, and local ones. The ADA adopts a definition with a history of acceptance dating from the 1973 Rehabilitation Act. The definition I used when I headed Ralph Nader's Disability Rights Center in the early 1980s was that a disabled person was anyone other than the mythical American—the 5-foot-10-inch, 160-pound, twenty-eight-year-old white male with no physical or mental impairments. Much of our public policy has been centered on improving society for that mythical American.

In my view the best book on disability is still Huxley's *Brave New World*. It was written as political satire in 1933, but I think we as a nation have gone to great lengths to make this satire a reality by segmenting society and relegating certain people to second-class status.

There was a time when left-handed schoolchildren had their left hands tied behind their backs and were forced to write right-handed. The reason for this was that it was cheaper to make tools exclusively for right-handed people than to make roughly 85 percent for right-handed and 15 percent for left-handed people.

Henry Ford said that you could buy his cars in any color you wanted as long as it was black. He made them only in black for the mythical American; others were excluded. In the same way, people who were different—whether disabled, older, younger, or whatever—were excluded from participation in society.

We can see an example of how far we have come toward a Brave New World in the Metro system here in Washington, D.C. The courts held (because of a lawsuit brought by a good friend of mine) that all Metro stations must be accessible to disabled people and therefore elevators must be installed. Disabled and elderly people are supposed to use the elevators and everybody else is supposed to use the escalators.

There are 60,000 reported injuries a year on escalators. They are very dangerous. They are especially dangerous for young children in the wintertime; the Metro system has had to pay some large settlements to the families of children who have been killed when their clothes got caught in the escalators. If the D.C. Metro had been designed for all ages and not just for the mythical American, there would be escalators for fewer than half the people in our society, and elevators for the majority—including those over forty-five with sight problems; the elderly; adults with strollers; people using wheelchairs; those with mobility or grip impairments; those who are subject to seizures. The list goes on and on.

The disability rights movement is a very recent development—less than twenty-five years old. We have made extraordinary progress in this relatively brief period because our industrial system is changing rapidly. The assembly line, where the disabled do not fit in, is giving way to a computer society, where disabled people fit in very well.

One objection I have to the chapters by Weaver, Rosen, Oi, and Leonard is a failure to recognize the tremendous prejudice against disabled people that exists in our society. Incentive-based policies, which do away with disincentives and create effective incentives, would work well only if there were no prejudice. We know from many studies that some people feel uneasy around disabled people. Typically these are people who fear getting old themselves. Disabled people remind them that they will get older and could become disabled. People who have not come to terms with death find disabled people frightening. We need to deal with the prejudice, myths, and misconceptions that arise from these fears.

Two other specific concerns should be raised. First, the chapters by Weaver and Rosen focus on the cost of accommodating people with disabilities. They do not factor in the tremendous costs of excluding disabled people from society. In a recent study, Monroe Berkowitz, Dale Hanks, and Stanley Portney found that the United States pays roughly $170 billion a year to exclude the disabled.[1] If we provide the accommodations that would make society accessible to all, we will save money in the long run. Nevertheless, I agree with Weaver and Rosen that important public policy issues surround the questions, Who should bear the cost of accommodation? and Should accommodation be an employer-mandate with no direct public financing?

Second, Oi's presentation of paratransit is problematic. Paratransit has really never worked as a method of mainstreaming people with disabilities. It is too labor-intensive; a small van and a driver can transport only a couple of people. The report cited by Oi grossly underestimated the number of wheelchair users in the country. Roughly 1 percent of our population are full-time users of wheelchairs. Accessible transportation can take the pressure off paratransit, which is needed for people who have health problems or older people who cannot use the regular transit system.

PART TWO

Economic and Legal Problems in Defining and Deciding Disability

5

In Search of the Disabled under the Americans with Disabilities Act

Jerry L. Mashaw

My task is to discuss the process for determining who is "disabled" pursuant to the employment discrimination provisions of the Americans with Disabilities Act (referred to as the ADA).[1] Given my prior work on disability determination in the context of the social security disability insurance (DI) program,[2] my initial reaction to the legislation was skeptical at best. Disability is, after all, a highly complex concept, constructed from the interaction of medical, vocational, and personal psychological characteristics. The interpretation of who is disabled under the Social Security Act engages, and routinely baffles, an army of well-trained adjudicators at the Social Security Administration and its affiliated state agencies. SSA's disability determinations annually produce between 250,000 and 500,000 formal administrative adjudications and account for 7,000 to 20,000 new filings in federal district courts. Extensive administrative, congressional, and scholarly attention to the social security disability adjudicatory process over several decades has failed to make the standard for disability determination significantly less opaque or to reduce the level of legal contest that surrounds the program. Moreover, resolving the admitted difficulties of social security disability determination has eluded SSA policy makers in a system blessed with considerable managerial talent that has been applied diligently to the centralized control of erroneous, untimely, and inefficient decision making.[3]

It is hardly surprising, then, that when I learned Congress was proposing to put a similar question, Who is disabled?, to millions of disappointed job applicants or employees and to their employers in a decentralized scheme relying on the enforcement machinery of the 1964 Civil Rights Act, my reaction was that Congress must have taken leave of its collective senses. The Civil Rights Act of 1964, after all, has produced a rather spectacular amount of litigation in its twenty-five-year history. A brief look at the U.S. Code Annotated reveals that the reported cases interpreting just the principal remedial section of that act, 42 U.S.C. section 2000e-5, require 750 pages in the USCA for the standard one-sentence descriptions of their holdings. And this level of litigation has been generated under a statute that leaves little doubt about who is a covered beneficiary. Adding the highly problematic category of disabled to an already highly litigious arena was not self-evidently sensible. The Congress that in 1988 set up a commission to study ways of easing the burdens on the federal judiciary now seemed intent on burying that same judiciary in an enormous new class of lawsuits. This same Congress also seemed to have forgotten that, however well the federal judiciary has done with discrimination litigation, its understanding of disability cases has never drawn the praise of informed commentators.[4]

Upon further reflection, however, I am not at all certain that my initial reactions were sound. Those reactions may have been based on an inappropriate analogy. I am most familiar with a system in which the interpretative conundrums surrounding the concept of disability are made manageable, and then only barely so, by attending continuously, through a centralized process, to the regularization of disability decison making. I therefore leapt quite easily to the conclusion that a highly decentralized process for applying a similar concept was likely to produce chaos, and costly chaos at that. But in doing so I forgot that, in interpretation, context often overwhelms text.

To understand the difficulties of determining who is disabled under the Americans with Disabilities Act, we must consider both the particular purposes of that statute and its different operational context when compared with the DI program. The analytical difficulties besetting DI adjudication may be simplified radically by a scheme whose purpose is inclusion rather than technical differentiation. Moreover, the operational realities of policing disability discrimination may be very different from the operational realities of a disability benefits program. The way in which the disabled see themselves and the ways in which others see them will be shaped significantly by the incentive structures surrounding each act of conceptualization. In the ADA context, those incentive structures may be quite different from the ones surrounding the social security disability programs. Finally, the criteria for determining whether the decision process for applying the statute is effective or efficient may also be shaped by public purpose and policy context. It is not clear that the paramount values of accuracy, timeliness, and efficiency surrounding a bureaucratic decision process like that of the Social Security Administration are appropriate for a nondiscrimination statute that hopes to promote widespread social change.

Let me take up these topics in turn: first, the analytical difficulties in determining disability pursuant to the Americans with Disabilities Act; second, a brief, speculative exploration of the operational context within which disability determination will be made under the ADA; finally, some even briefer reflections on the appropriate criteria to be applied to the determination process this statute will establish.

The Analytics of Disability Determination

Using the concept "disabled" to define a protected class for purposes of antidiscrimination law poses an immediate puzzle. Disability is, after all, a functional concept. Our experience with protected classes under prior antidiscrimination statutes—persons discriminated against because of race, sex, religion, national origin, or age—is with efforts to eliminate invidious decisional criteria that are in no systematic or demonstrable fashion related to the functional capacity of individuals. What then could it mean to discriminate against the disabled in employment, when determining that persons are disabled implies precisely the determination that they are functionally incapable of carrying out one or another task?

Under the Americans with Disabilities Act, discrimination against the disabled could take a number of forms—forms that, on examination, are not so different from our standard interpretations of employment discrimination in other contexts. First, employers might exclude the handicapped from job opportunities or benefits when their disability has no rational relationship to their ability to perform a job. The exclusion of nondrinking alcoholics or all HIV-positive individuals might be examples. This is straightforwardly disparate treatment of the disabled and poses no peculiar analytical problem.

Second, employers might use general criteria that have diverse and unintended impacts on the disabled and are unncessary in determining an individual's ability to do a job. The use of general written tests, for example, might exclude most of the blind from applying for many jobs that require no ability to write, or even to read, ordinary characters.

Third, and similar to the case of seemingly neutral criteria with unforeseen impacts, is the rejection of disabled persons because of barriers to job performance that could be ameliorated by the employer. Such ameliorations might be as simple as providing a telephone amplifier or as complex and expensive as the architectural redesign of a plant. Under the ADA, the failure to make reasonable accommodations to deal with worker handicaps for those otherwise qualified to perform jobs is itself discrimination. The act thus contains within it an affirmative action duty that puts some strain on the ordinary meaning of "discrimination," but affirmative action requirements are certainly not unfamiliar in discrimination law generally.

Note first, then, that while discrimination against the disabled may initially seem a peculiar concept, its occurrences are not far removed from our conventional understanding of antidiscrimination policy. The real difference between this class of persons and those covered under prior law is our intuition that here the discrimination issue will resolve itself more frequently into one of determining the reasonableness of the employer's job criteria or efforts at accommodation. In other words, we are likely to believe that in most cases the employer can present a superficially plausible explanation for excluding an employee or job applicant when that exclusion is based on disability, but that most exclusions on the basis of race, sex, religion, or national origin would be extremely difficult to justify. Thus, questions of who is an "otherwise qualified" individual or what accommodations by an employer can be made without "undue hardship" are likely to raise difficult issues of judgment under this statute. But these issues go to questions other than the basic determination of who is disabled for purposes of the act's coverage.

To see whether difficult issues of interpretation or application surround the concept of disability, we must delve more particularly into how that term is used in the ADA. According to the statute,

> The term "disability" means, with respect to an individual—
> (A) a physical or mental impairment that substantially limits one or more of the major life activities of such individual; (B) a record of such an impairment; or (C) being regarded as having such an impairment.[5]

At first blush this definition certainly seems capable of causing problems. Phrases like "substantially limits" or "major life activities" are hardly self-interpreting. Indeed, they might be thought so highly judgmental or contextual as to defy abstract explication. How are employees and employers to grapple with such terms? Is not the disability standard in the ADA both an open invitation to protracted litigation and a source of costly legal uncertainty in other ways as well?

Perhaps, but an affirmative conclusion now seems much too hasty. There are sources of guidance available that may make the interpretive task more manageable. First, this statutory definition has a substantial history. It is the same as that used in section 504 of the Rehabilitation Act of 1973, as amended, a definition that has also been incorporated into a large number of state nondiscrimination statutes.[6] Interpreters may, in cases of doubt, refer to administrative guidelines and regulations as well as judicial decisions under these prior statutes. Second, employers, employees, and courts can expect to obtain further elaboration of the disability concept soon after the passage of the ADA. The Equal Employment Opportunity Commission has the responsibility for issuing regulations to elaborate this definition. Moreover, the legislative history of the ADA suggests that its sponsors contemplate regulations similar to those adopted pursuant to the Fair Housing Amendments Act of 1988 and those adopted by the Department of Health Education and Welfare pursuant to the Rehabilitation Act of 1973.[7]

These existing regulations take a broadly inclusive approach.[8] They specify, for example, that any physiological, mental, or psychological disorder or condition (other than addiction caused by current illegal use of a controlled substance) will qualify as a "physical or mental impairment," provided that such disorder or condition has a substantial effect on a "major life activity." The latter term is then defined expansively to include "functions such as caring for oneself, performing manual tasks, walking, seeing, hearing, speaking, breathing, learning and working." "Substantially limits" is not a defined term in the regulations, and, as we shall see, it has proved troublesome in court.

Nevertheless, the existing regulations suggest an expansive policy here as well. In explicating the third clause of the statutory disability definition, for example, the HUD regulations pursuant to the Rehabilitation Act of 1973 make clear that HIV infection is a covered impairment because it substantially limits a major life activity, even if that limitation is only as "a result of the attitudes of others toward" the infected individual.

The concept of disability thus has a function in the Americans with Disabilities Act radically different from its function in the social security disability insurance program. Rather than serving as a conclusion that has the legal effect of entitling a person to benefits, like in the DI program, here the classification merely defines a protected class that may or may not in particular instances have any right to a legal remedy. It is a categorization that begins the analysis of a case rather than concluding it. The ADA's prudential limitations are built into other key concepts in the statute, such as whether persons are "otherwise qualified," whether "reasonable accommodation" has been made to an impairment, or whether accommodation to an applicant's or employee's disabilities would impose "undue hardship" on the operation of the business. In the DI context, by contrast, congressional concern with overgenerous provision of disability benefits is built into the definition of disability itself. Hence, whereas the DI system forces all the difficult trade-offs between social compassion and work incentives into the interpretation and application of the concept of disability, the ADA shifts similar issues of balancing employee needs against employer costs elsewhere in the analysis of a case.

A recent decision of the U.S. Supreme Court interpreting section 504 of the Rehabilitation Act of 1973 supports this interpretive approach. That case, *School Board of Nassau County, Florida v. Arline,* involved the removal of a public school teacher who suffered from recurrent bouts of tuberculosis.[9] The district court held that Arline was not a handicapped person under the terms of the 1973 act. In its view, the school board had removed her not because of a handicap (the board had never suggested that she could not do her job), but because of her contagiousness, which posed a danger to others. The Court of Appeals reversed this decision, holding that persons with contagious diseases are within the coverage of section 504. It remanded the case, however, for further findings as to whether the risks of infection precluded Ms. Arline from being otherwise qualified for her job, even after making some reasonable accommodation for her condition.

The Supreme Court affirmed in an opinion that strongly reinforced the inclusiveness of the concept of disability or impairment in the rehabilitation act. The Court also made clear that the question of whether Arline was otherwise qualified for her job would entail specific findings of fact concerning the nature, duration, and severity of the risk she posed to others because of the contagiousness of her disease.

This is not to say that every determination of whether a worker or potential worker is disabled or impaired or handicapped can be finessed by referring the issue to some other crucial term in the statute. Nor is the Congress's implicit instruction "when in doubt be inclusive" always enough to resolve difficult cases. A prime example of the difficulties that can emerge is the case of *E. E. Black, Ltd. v. Marshall.*[10] Plaintiff Crosby was dismissed from a carpentry apprentice program after a medical examination determined he had a congenital back abnormality predisposing him to back disease or injury. At the time of his employment Crosby was asymptomatic. Indeed, he was in strapping good health and testified that because of his size he was often assigned to some of the heavier jobs on the building site.

The administrative law judge who heard Crosby's complaint

found that he was not covered by the Rehabilitation Act of 1973. Although he had a physical condition or abnormality, the ALJ viewed that abnormality as barring him only from certain jobs involving heavy lifting. It was therefore not "substantially limiting" a "major life activity." The administrative law judge applied a test something like the DI concept of "slight impairment." Because most jobs remained open to Crosby, he was not "substantially limited."

On appeal to the Department of Labor, the ALJ's determination was reversed. The assistant secretary took the opposite tack, holding that anyone who had an abnormality that was the basis for a negative employment determination had been discriminated against under the act, unless the individual was not capable of performing the federal contractor's valid job requirements. Under this view, proof of substantial limitation on a major life activity is satisfied by the mere fact of some limitation on employment being premised on an existing physical or mental impairment.

E. E. Black appealed the assistant secretary's decision to a district court in Hawaii. That court rejected both the administrative law judge's and the assistant secretary of labor's approaches. The court thought that analyzing employability generally, as the ALJ had done, undercut the purposes of the act by failing to consider a person's training or vocational goals. But the court also thought that the assistant secretary's interpretation ignored the act's requirement that the impairment have some substantial effect on a major life activity. It reached the same ultimate result as the assistant secretary, but only after considering both the range of jobs that would be denied Crosby should all employers take the position that E. E. Black took, and the degree to which those decisions would substantially affect Crosby, given his particular qualifications and aspirations.

While the *E. E. Black* decision may be perfectly sensible as an act of judicial statutory interpretation, the court's approach apparently envisages a rather messy case-by-case determination by employers. Decision makers are invited, indeed required, to take into account not only the employers' requirements and situation but also the specific vocational aptitudes and aspirations of particular applicants or employees. Making such determinations has direct costs for everyone involved, and erroneous determinations because of legal uncertainties may impose further costs resulting from over- and undercompliance. For this reason the EEOC might be well-advised in developing regulations for the ADA to include something like the assistant secretary of labor's view in *E. E. Black* as a definition of "substantially affects."

The Effect of Operational Context

Detailed regulatory action by the EEOC, then, is one remedy for the residual legal uncertainties surrounding the disability standard in the ADA. But regulations, however detailed, will never resolve all the interpretive issues of this (or any) statute in advance. There are many marginal categories that raise difficult, even daunting problems. Is it a disability to be left-handed? obese? short? Must the condition be immutable? Can it be self-induced? All of these and many other issues are potentially available in the context of discrimination claims. And indeed, some of them have arisen.[11] The question is, How serious, widespread, and costly are these legal uncertainties likely to be?

Litigation experience under the Rehabilitation Act of 1973 and similar state statutes suggests that the answer is, not very. Although these statutes have been in force for a decade or longer, there is surprisingly little jurisprudence on the meaning of disability. A plausible interpretation of this lack of reported cases is that potential litigants are generally neither uncertain nor in dispute about the question of who is disabled under these statutes. For if litigants have the same view of the case, there is no reason for them to litigate rather than settle.[12]

The plausibility of this explanation is enhanced by considering the quite different contexts of litigation under the DI program and the ADA. Disability insurance claimants tend to be quite severely impaired, unemployed, and without access to alternative programs for income support. Both their own experience and their physicians' advice have convinced them they cannot work. Once they have overcome what is for many a strong reticence to apply for benefits, the suggestion by the Social Security Administration that they are not disabled seems to them clearly erroneous, if not perverse. This combination of economic pressure and psychological commitment, when added to a stringent disability standard, explains much of the continuous controversy that surrounds disability determinations in social security. Because of the structure of that statute, disputes focus on one issue—Is the claimant disabled? And the outcome is all (a full social security retirement pension) or nothing.

The operational context of the Americans with Disabilities Act should be quite different. First, most cases are likely to entail no dispute about whether the claiming party has some disabling condition. The existence of that condition will have been the predicate upon which the employer has refused either to begin or continue employment or to give the claimant access to some other benefits or advancement possibilities that are attached to a job. Disputes will much more naturally focus on whether the complaining party is otherwise qualified or whether the requested accommodation to the claimant's condition is a reasonable one.

Second, persons with marginal impairments, and therefore with a range of other job prospects, may find it quite uninviting to press a claim of disability discrimination. The payoff is employment or retention by an employer who obviously is not well disposed toward the employee. And because being disabled or handicapped carries negative connotations for most persons, they are unlikely to assign lightly or to insist on this categorization of themselves. Here the game is often not worth the candle.

The strategic situation also militates against making the condition of being disabled a battleground between employer and employee. Employees who must fight to demonstrate disability must also bear the burden of demonstrating that they are otherwise qualified for the employment. Proving the first will in most cases go some way toward disproving the second. By the same token, the employer who has used an abnormality as a grounds for dismissal or for refusal of employment will often look rather silly claiming that this condition is not an impairment, disability, or handicap.

This explanation is, of course, not entirely convincing. Direct actions have been available under the 1973 Rehabilitation Act only against a limited class of defendants. The broader reach and higher visibility of the ADA may tap wellsprings of litigiousness not approached by either the 1973 act or prior state laws. I cannot be certain that the absence of litigiousness, which I have ascribed to the strategic context of the application of the disability concept in the federal rehabilitation and state antidiscrimination acts, may not instead have resulted from the limitations of prior federal protections and from the lack of general knowledge of state law remedies.

Assessing the Disability Determination Process

What then can we expect in the application of the disability standard of the Americans with Disabilities Act? My own prediction is that the level of dispute over this issue is likely to be low and that most disputes will be resolved fairly easily in terms of the general purpose of the statue to be inclusive in its coverage. Nevertheless, disputes will arise, and in a decentralized system relying on both self-application and decision making by multiple district and circuit courts we can expect some variance in results. The question is whether a different decision process should be designed to lend uniformity to determinations of disability under the statute. Perhaps some machinery for certifying disabilities could be established, making claims of discrimination available only to those previously determined to be suffering from a disability. It might even be possible to rely on some broadened variant of current local processes for assigning handicapped parking permits.

My own tentative view is that further regulatory action to promote uniformity is probably unnecessary and perhaps dysfunctional. The ADA is not like the social security DI program, in which there is a national commitment to uniform administration of statutorily defined public entitlements. It is instead an attempt to reshape the public consciousness with respect to disabled persons and to promote a process of accommodation that better integrates such persons into a number of areas of public life, including employment. A decentralized process of the sort entailed here—beginning with employer and employee determinations, proceeding through attempts at amelioration through EEOC intervention, and ultimately resorting to district court litigation—provides multiple opportunities for the shaping and reshaping of our ideas about the position of the disabled in the workplace. Consistency may not have a very high priority in this process.

Hence, while I agree with much of Carolyn Weaver's analysis of the potentially disparate impacts of the ADA on subsets both of the disabled and of private employers, I do not share her dismay at these inevitable inconsistencies.[13] All laws have disparate impacts; all rights are differentially useful or valuable to their holders and differentially onerous to those who have corresponding duties. The serious concerns, it seems to me, are the others that she suggests: that the ADA employs potentially unfair taxation to provide in-kind benefits, which a deficit-happy Congress does not want to fund through the budget process; that the costs of providing those benefits may exceed their value.

Neither of these problems, however, is necessarily a function of legal uncertainty surrounding the concept of disability in the statute. Maldistribution of the tax burden will occur, if it does, because of varying situations, not because of legal uncertainty. And ex ante, it is virtually impossible to predict whether legal uncertainty will induce under- or overcompliance with the law.[14] It is only in the case of high planning, dispute, and litigation costs that legal uncertainty produces unambiguously negative economic consequences. But as I have said, I do not see a strong reason to expect that result.

Moreover, we might do better to think about this new initiative in a dynamic rather than static fashion. Numerous employers who have pioneered in employing the handicapped or who have made accommodations for the disabled in compliance with prior federal and state requirements report either negligible costs or efficiency gains from those efforts.[15]

It may even be the case that a high initial level of legal uncertainty, including disputes, negotiation, and litigation, is a necessary device for raising public consciousness about the plight of those who have some impairment—but nonetheless a determination to be self-supporting—and for harnessing the energies of the private sector to the task of ameliorating the disadvantaged position of the disabled. If so, perhaps the "creative destruction" of whatever legal uncertainty is generated by the ADA should be welcomed rather than condemned. For this statute, after all, does not give people incentives to take tickets out of the work force. It instead gives legal leverage to people who want, indeed demand, to work. In determining the set of property rights we all hold in a market economy, it seems sensible to accept some transaction costs, even error costs, for giving people the right to fight for an opportunity to be productive.

6
Measuring and Deciding Disability

Donald O. Parsons

The problem of deciding who is disabled has seriously limited public and private efforts to provide insurance coverage against earnings losses due to disability. The nature of the problem is twofold. First, it is difficult to disentangle the economic losses resulting from physical or mental impairments and those resulting from labor market conditions, and indeed these losses are likely to be interactive in nature. Second, judges seem unable to ignore the economic hardships of the individuals affected, no matter what a reasonable interpretation of the insurance language might indicate. During the Great Depression, for example, judges were either unable or unwilling to make the important distinction between earnings losses due to disability and earnings losses due to the extreme labor market conditions of the time, and many disability insurers went bankrupt; the private market for disability insurance all but collapsed.[1]

A somewhat ironic consequence of this private market failure was the retardation of public efforts to insure against earnings losses due to disability. Only one type of disability impairment, blindness, was covered by the Social Security Act of 1935. Other physical conditions may have been equally threatening to economic well-being, but blindness was viewed as insurable because it imposed systematically large losses on those afflicted and was physically verifiable. Social security did not begin paying cash benefits to other types of disabled workers until 1957 and then only to people fifty years of age and older.

This chapter explores the problem of measuring and deciding disability, as revealed by our experience with the social security disability insurance (DI) program, and considers alternative approaches to improving the disability adjudication system.

The Disability Measurement Problem

The insurance objective inherent in DI is compensation to individuals for earnings losses due to the onset of disability. Earnings losses may arise for a variety of reasons, some of which are not insurable, and it is therefore essential to identify their cause. Stubborn measurement problems exist, unfortunately, in assessing impairments and any resulting loss of earnings. Even when the impairment is known, estimating the extent of earnings losses due to the impairment is fraught with difficulties. Seemingly equivalent impairments account for wide differences in earnings losses. Individual effects, including motivation and energy, are critical and largely unmeasured. Almost no impairment is so severe that someone somewhere in the economy is not earning a satisfactory living while suffering from it. The apparently random element in this relationship introduces considerable uncertainty into the loss assessment and provides generous room for dispute.

The measurement problem is compounded by the disincentives to accurately self-report the severity of disability inherent in any social insurance program. People are paid to work for a very good reason; most would rather be doing something else, whether indulging in leisure activities, caring for their families, or pursuing another line of work that is personally more satisfying but less remunerative. In this environment self-reporting of health conditions or disability-related earnings losses cannot be relied upon as the basis for compensation. The loss of this critical input—unbiased self-assessment—is a serious one. Medical evaluations depend in an important way on the cooperation of the individual, as do economic assessments of the source of earnings losses.[2]

The human factor is likely to affect judicial behavior as well as the behavior of applicants. Again recall the Depression-era difficulties that judges found in adjudicating disputes over private disability contracts. Cases that are either factually ambiguous or highly emotional are likely to be determined primarily by judicial preference. Those who may be ineligible under the law are likely to be in unfortunate circumstances, both relatively poor and in relatively bad health, and they are obvious targets for sympathy. On the other hand, for a given health condition they are likely to be the individuals with the least determination and drive. How a judge views such cases will vary from judge to judge.

Lessons from the DI Program

The depth of the disability measurement problem in a social insurance framework is well illustrated by the current social security program. A major concession to the assessment problem is the limitation of coverage in the social security system to those with a total disability. The official definition of a compensable disability, which has changed only modestly over time, is

> the inability to engage in any substantial gainful activity by reason of any medically determinable physical or mental impairment which can be expected to result in death or which has lasted or can be expected to last, for a continuous period of not less than 12 months.[3]

The individual is on his or her own to cover earnings losses resulting from more modest disibilities.

An elaborate administrative structure is in place to determine eligibility under this criterion. The initial determination of eligibility is formally conducted by state agencies on behalf of the Social Security Administration (SSA). If negative, this decision is then subject to an elaborate appeal system. The denied applicant may first petition for a reconsideration by the state agency. If the applicant is again denied, he has the right to appeal to an administrative law judge (ALJ). If that fails, he has the right to appeal to SSA's Appeals Council. Failing that he may proceed to the U.S. court system at the district court level.

A number of studies have documented the imperfections in the disability screening technology employed by SSA. R. T. Smith and A. M. Lilienfeld, for example, undertook a study in which they sent back for review by SSA's Bureau of Disability Insurance (BDI) a sample of 250 allowance decisions and a like number of denial decisions (248).[4] The distribution of initial determinations and review determinations are reported in table 6–1, panel A. Approximately 20 percent of the cases allowed by BDI had been denied initially (21.2 percent), and 20 percent of the cases denied by BDI had been allowed initially (22.5 percent).

In a thorough evaluation of SSA's disability screen, undertaken by Saad Nagi and associates, a sample of 2,454 initial disability determinations was drawn from three geographic areas (Louisiana, Minnesota, and Ohio).[5] The individuals were intensively examined by a clinical team that included a physician, a social worker, a psychologist, an occupational therapist, and a vocational counselor. The cases were then sent back for redetermination by SSA with the additional information accumulated by the clinical team included in the file. The differences between the initial determinations and the redeterminations are reported in table 6–1, panel B. Review of the files with the additional information provided by the study team led to 8.2 percent of the initial allowances being changed to denials and 20.8 percent of the denials being changed to allowances.

Of particular interest is the relationship between the clinical

74

TABLE 6–1
REASSESSMENTS OF INITIAL SOCIAL SECURITY DETERMINATIONS
(percent)

*Panel A: Bureau of Disability Insurance Review One Year
after Initial Determination[a]*

	Initial Determination	
	Allowance	Denial
BDI Assessment		
Allowance	78.8	21.1
Denial	22.5	77.5

Panel B: SSA Redetermination Decisions by Initial Determination[b]

	Redeterminations	
Initial Determination	Allowance	Denial
Allowance	91.8	8.2
Denial	20.8	79.2

NOTE: Allowances and denials may not total 100 percent because of rounding.
a. The sample sizes are 250 initial allowances and 248 initial denials.
b. Weighted sample size is 2,454: 1,434 initial allowances and 1,020 initial denials.
SOURCES: For panel A, see R. T. Smith and A. M. Lilienfeld, "The Social Security Disability Program: An Evaluation Study," U.S. Social Security Administration, Office of Research and Statistics Research Report, no. 39, 1971. For panel B, see Saad Z. Nagi, *Disability and Rehabilitation: Legal, Clinical, and Self-Concepts and Measurement* (Columbus: Ohio State University Press, 1969), p. 60.

team's assessment of functional limitation and the final SSA determination. Recall again that SSA had access to the clinical team's information. The variations in SSA's final determinations resulting from the clinical team's assessment of work capacity are reported in table 6–2. There is a general concordance of judgments between the two decision groups. The SSA allowance rate increases continuously from 0 percent to 75.2 percent as the clinical team's assessment of disability severity increases. Nonetheless the differences in the two assessments are not small. Of the 1,019 applicants judged not fit for work by the clinical team, one-quarter (24.8 percent) were judged by SSA to be capable of engaging in substantial gainful employment and therefore ineligible for benefits. Of the three groups judged by the clinical team to be fit for full-time work under normal conditions, 117 out of 435 (26.9 percent) were granted disability status by SSA.

TABLE 6-2
FINAL SSA DETERMINATIONS OF DISABILITY IN RELATION TO CLINICAL TEAM'S EVALUATIONS OF WORK CAPACITY

	Final Determination					
	Allowance		Denial		Total	
Work Capacity	Number	%	Number	%	Number	%
Fit for work under normal conditions	0	0.0	9	100.0	9	100.0
Fit for specific jobs, including former job, under normal conditions	23	13.9	142	86.1	165	100.0
Fit for specific jobs, excluding former job under normal conditions	94	36.0	167	64.0	261	100.0
Fit for work under special conditions	92	50.5	90	49.5	182	100.0
Can work part-time under normal conditions	82	49.4	84	50.6	166	100.0
Can work under sheltered conditions	134	60.6	87	39.4	221	100.0
Can work at home only	66	69.5	29	30.5	95	100.0
Not fit for work	1,019	75.2	336	24.8	1,355	100.0
Total	1,510	61.5	944	38.5	2,454	100.0

SOURCE: See source for panel A, table 6–1.

Whether the clinical team or SSA was right in its decisions is unclear; the ideal may have eluded both groups. It is clear that the ambiguities are substantial in both measuring mental and physical impairments and estimating their labor market implications.

The intrinsic measurement problem is compounded by the heterogeneity of judicial preferences. Studies of the ALJ stage of the disability appeal structure reveal the extent to which the judicial preference problem plagues the adjudication process.[6] The ALJ stage is by design independent of SSA. The issue under review is an appeal of a negative SSA decision; ALJs are given the task of determining whether SSA's denial of eligibility is valid. Theoretically ALJs are committed to following SSA regulations for decision making, but practically they have little incentive to do so. The possibility exists that they will decide cases solely on the basis of their own redistributive preferences and not on the basis of social security regulations.

The variability of ALJ preferences and decisions is reflected in data reported by Dixon on reversal rates by ALJs: that is, the proportion of cases (or appealed denial decisions) that are reversed and allowed benefits by the ALJ.[7] The distribution of average reversal rates in fiscal year 1971 by individual (full-time) ALJs, each of whom typically had a large number of cases, is reported in table 6–3, column 1. The striking aspect of the reversal distribution is its dispersion; more than 10 percent of the examiners had reversal rates of 20 to 30 percent, while 3 percent had rates between 70 and 80 percent. (The average reversal rate was 46 percent.) The median examiner reviewed about 150 cases, so that under the hypothesis that each examiner had an equal probability (0.46) of reversing an individual case, the probability of deviating substantially from a reversal rate of 46 percent is small. An ideal distribution constructed under this hypothesis is reported in column 2 for comparison purposes. In a sample of 252 examiners each with 150 cases, for example, the probability that any judge's average reversal rate is as low as 20–30 percent or as high as 70–80 percent is approximately zero. Thirty-four examiners were in fact found in those extreme categories. It appears that some examiners felt few disabled workers warranted income maintenance, while others felt few disabled claimants did not warrant such support. A claimant would surely prefer his or her case to be heard by an examiner with an 80 percent reversal rate rather than a 20 percent rate. Claimants have no such choice, of course; from their vantage point the ALJ process is essentially random, independent of the facts of the case.[8]

The average judicial tendency toward leniency is more difficult to assess. Certainly the history of interaction between the federal

TABLE 6–3

Reversal Rates among Full-time SSA ALJs, FY 1971

(percent)

Examiner Reversal Rate	Actual Frequency of Examiners (1)	Hypothetical Frequency of Examiners (2)
0–20	0.00	0.00
21–25	3.17	0.00
26–30	7.14	0.01
31–35	8.73	0.48
36–40	11.51	8.36
41–45	19.05	36.37
46–50	16.67	41.43
51–55	13.49	12.36
56–60	7.94	0.97
61–65	4.76	0.02
66–70	4.37	0.00
71–75	2.38	0.00
76–80	0.79	0.00
81–100	0.01	0.00

Note: The total number of examiners was 252. The hypothetical distribution was constructed from a normal approximation to the binomial distribution generated under the assumption that each examiner reviewed 150 cases, with an expected reversal rate of 0.46 for each case.

Source: R. G. Dixon, Jr., *Social Security Disability and Mass Justice: A Problem in Welfare Adjudication* (New York: Praeger Publishers, Inc., 1973), p. 78.

courts and the legislature is one of lenient judicial interpretation of statutes, followed by more stringent statutory requirements to offset the judicial tendencies. The consequences of ruling a marginally disabled worker ineligible is immediate, predictable, and harsh. The costs of an allowance are less direct and, one would speculate, less discomforting to the decision maker, particularly if he is short-sighted. The efficient judge must have the time and inclination to appreciate the budgetary implications of his decision and the possible long-run consequences of a lenient ruling on the acceptance of the program by legislators and the electorate. Few judges are likely to have the time or inclination to consider such broad policy questions.

Implications for Program Performance

Measurement and implementation problems of the magnitude of those plaguing the social security disability program are likely to lead to an insurance program that is both inefficient in its delivery and costly to provide.[9] This appears to be the case with DI. The benefits received by an eligible disability applicant are not generous. The basic benefit rule is the same one used for age sixty-five retirees; in the case of disability, however, these benefits are subject to a very large deductible. Benefits are paid only after the individual has been out of the labor force for five months; the first five months of total disability are not covered. This provision no doubt makes the program less attractive to marginal claimants, but only at the cost of making the target group of severely impaired workers suffer significant earnings losses.

The expected benefits of a disability application are much lower than even these restrictions would suggest. More than 50 percent of disability applicants (individuals who claim to be totally disabled) are initially rejected by SSA, as persons considered able to engage in substantial gainful employment. The magnitude of this rejection rate attests to the randomness of the process. Individuals find it hard to predict whether they are covered or not; otherwise they would not have made the substantial investment of lost work time required to enter this lottery. A large percentage of denied applicants appeals SSA's eligibility determination, often spending years in the appeal process. The determination of eligibility often requires the individual to make a very substantial investment of time.[10]

The aggregate economic cost of operating the disability system in this world of incomplete information is high.[11] Trends in labor force participation suggest that the work disincentive effects of the program are large. Adult males in the United States are much less likely to be in the labor force than they were thirty years ago. Between 1948 and 1983 the percentage of males sixteen years of age or older who neither had jobs nor sought them rose by 10.2 percentage points, or one-tenth of the adult male population. The nonparticipation of black adult males has risen yet more sharply, at a rate 50 percent higher than the population's as a whole. This rapid rise in the rate of adult male nonparticipation is not simply the result of the young seeking additional education or the old choosing early retirement. The percentage of adult males forty-five to fifty-four years of age who are out of the labor force rose from 4.2 percent to 8.8 percent between 1948 and 1983; the rate of nonparticipation of those thirty-five to forty-four doubled, from 2.1 percent to 4.8 percent. This large increase in the nonparticipation rate of prime-age males matched closely the rapid expansion of the DI program since its inception in the mid-1950s. In 1987 approximately 2.7 million disabled workers were receiving social security disability benefits totaling $20.5 billion. Related supplemental security income benefits, federal and state, totaled an additional $9.5 billion.

79

Econometric studies indicate that the disability application and award (or allowance) rate and the labor force participation rates of older workers are in fact sensitive to the generosity of disability benefit levels. Internal studies by SSA, particularly those of Morde-chai Lando and associates, have estimated a significant sensitivity of disability applications to labor market conditions and the generosity of benefits.[12] My own work indicates that a 10 percent increase in the ratio of benefit levels to market wages will induce a 5 to 10 percent increase in male labor force nonparticipation.[13] Jonathan Leonard, in a careful review of the literature, suggests that a good estimate might be that a 10 percent increase in the replacement ratio (the ratio of annual benefits to gross predisability annual earnings) will induce a 5 percent increase in labor force nonparticipation.[14]

A program in which eligibility for benefits is uncertain and eligibility determination contentious is likely to be inimical to rehabil-itation and indeed to self-help of any type. Disability recipients, for example, do little fine tuning of work effort to take advantage of existing possibilities to augment their incomes.[15] The "substantial gainful activity" level, the earnings level above which SSA considers the individual capable of gainful activity, appears to play quite a different role in this program from that of the earnings test in the social security retirement program. In the retirement program many individuals earn slightly less than the lower bound of the earnings test. The work of Paula Franklin and her associates in the mid-1970s suggests that this is not the case with the disability program and the SGA level.[16] Most recipients, perhaps 90 percent, simply do not work or at least do not work at jobs for which income is reported. Even among the small number of disability recipients who do work, the SGA level appears to have no unusual significance.

This is not surprising, given the many uncertainties that exist in the recipient's mind about current and future eligibility for disability benefits. Eligibility for disability is not easily secured and is at least in part a random event. How does an individual know what will get him or her into trouble? Just as an individual may pass up a legitimate tax deduction that could trigger an IRS audit, the recipient may well find it prudent not to raise questions about his or her fundamental eligibility for disability benefits by working. Policies may change in a future budget crunch, and valuable benefit entitlements may be lost as a result of recent work activity.

Directions for Improvement

Perhaps I should stress first what is *not* a solution to the measurement and implementation problems of the disability system. The solution is not simply more stringent screening. More stringent screening of disability applicants will no doubt reduce the number of individuals who are healthy enough to work yet receive government payments for not working. Such a goal is surely desirable both from a budgetary standpoint and as a matter of simple justice to taxpayers and the working poor. A more stringent screen, however, will just as inevitably increase the number of truly disabled who are denied government assistance. Screening rigor will make the program less expensive, but at the same time will make the program less reliable for the target group of severely disabled. Improving the accuracy of disability-screening technology is obviously desirable, if the improvement comes at a sufficiently low price.

The judicial preference problem is perhaps the most amenable to system redesign. The variability of judicial outcomes at the ALJ stage, for example, is amenable to control through increasing the size of the court or tribunal. At present one ALJ hears and decides each case. Averaging over a number of judges could limit the impact of the most extreme, although at the cost of increasing the number of judges. Cheaper to implement, although perhaps less politically attractive, would be "tail-clipping," or eliminating [from the system] those ALJs with the most extreme records; if the distribution of judges is clipped appropriately at both tails, the average acceptance rate can be maintained.

A more radical reform would be to redesign the disability adjudication structure entirely, replacing judges at all levels (except federal court) with a mechanical decision rule. Eligibility could be based on a disability score, for example, which would depend on observable impairment and vocational attributes of the individual and the labor market. Such scores are now common case selection aids in situations as diverse as home loan approval and IRS audits. In such a system individuals could be spared the necessity of withdrawing from the world of work and suffering the indignities of the adjudication process to learn whether they are eligible for disability benefits.

Even this system would be imperfect, of course. Any mechanical system of determining disability eligibility will fail occasionally. Some individuals who are worthy of disability insurance will fall through the cracks, while some who are able to work at little personal cost will be judged eligible. To work effectively, appeals from the mechanical system to a human one must be severely limited. Designed appropriately, the mechanical system should take into account the entire factual basis for the disability claim. Further adjudication of such cases would probably be random, with eligibility decisions based primarily on judicial preferences rather than on the merits of the case. For deserving applicants who are rejected because of the inadequacies of the government's information base, other systems with richer information sets, such as family and community, should perhaps provide the needed income support. In the end the government must recognize its own limitations; it is a cumbersome mechanism for doing good works. The unfortunate truth is that some injustices are too costly to remedy and are better left to family and community to correct. Although their resources may be more limited, their information and understanding of a particular situation may be much superior.

Disability Legislation and Litigation

A Commentary by Michael J. Astrue

Donald O. Parsons and Jerry L. Mashaw have written fascinating chapters on an extremely important and difficult issue: defining and deciding who is disabled. It may be instructive to consider the experience of the Department of Health and Human Services (HHS) in this regard as it is so deeply involved in disability policy.

As general counsel of HHS, I must defend the most sued person in the universe—the secretary of Health and Human Services. About 98 percent of our litigation is disability litigation. Almost all of it turns on the question of whether someone is disabled. In recent years we have averaged at any given time about 50,000 cases in federal court. We have now managed to lower that figure to about 33,000—a small improvement. The department is constantly wrestling with the question of who is disabled.

The definition of disability changes from context to context. In an enormous department like HHS we have many offices dealing with disability, each using a different definition under a different federal program. One office may be litigating against a nursing home, defending the rights of an AIDS patient, while another office is arranging financial support for the developmentally disabled.

Another wrinkle in disability policy is the question of work. One of the stated goals of the Americans with Disabilities Act is to promote work among the disabled by protecting their rights in the workplace. In the social security disability insurance context, "working disabled" is an oxymoron—the very definition of disability under the law implies complete inability to work. This has been problematic in several ways and has led to a series of ad hoc adjustments and corrections that undermine the general definition. Defining and determining disability varies considerably from one context to another.

The history of disability policy is informative when we consider program reforms. Examples abound of how apparently simple ideas grew to become incredibly complex. When we first started paying DI benefits in 1956, we expected the program to be fairly small and fairly simple to administer. The general consensus held that disability could be determined in a very common-sense, nonadversarial way by using what became known as the Potter Stewart criterion, "I know it when I see it."

That has changed greatly over the years. We have an increasingly adversarial, legalistic system. What appeared fairly simple at the start now occupies the vast majority of the 40,000 pages of program instructions issued by the Social Security Administration. Asking a disability examiner at the state level to apply this complex system to a particular individual who has come into his or her office can be akin to handing the Internal Revenue Code to a group of bookkeepers untrained in tax law and saying, "File the corporate taxes for the Exxon Corporation."

My common-sense advice to those interested in disability policy is to be wary of apparently common-sense proposals. Give some thought to the operational problems the government will have in implementing those apparently simple ideas.

Lessons from the West German Approach to Disability Policy

A Commentary by Richard V. Burkhauser

Donald O. Parsons and Jerry L. Mashaw have provided provocative analyses on the problem of defining and deciding disability. I agree with much of what they have to say. The remarks that follow respond to some of their specific points and then consider disability policy in the Federal Republic of Germany. The incentive effects and costs of the German system, which includes employment quotas for handicapped workers, reveal much about the problems likely to arise with the highly regulatory approach to job protection taken by the Americans with Disabilities Act (ADA).

One implication of the Parsons chapter that deserves elaboration is that if the goal of government policy is to encourage people who have health conditions to remain in the work force, then it is very bad strategy to intervene after an individual has begun receiving disability benefits. This is simply too late to be effective, since by this time major investments in being totally disabled have already been made.

As Parsons makes clear in his discussion of the economic incentives of social security disability insurance (DI), a person who is work-imparied and wants to receive DI benefits must be unable to do any work in order to satisfy the substantial gainful activity criteria in the law. To increase one's chances in this lottery, the best strategy is to become as disabled as possible and not to work in the labor force for at least five months. Government policy must recognize that disability, or the movement toward disablement, is a dynamic process that can be affected by the design of government policy.

Although I have a number of reservations with the ADA, the law does have a positive attribute, which is that it seeks to keep people who become work-impaired on the job; through its job accommodation mandate it intervenes before these workers have left their jobs and become trapped in the disability system. This is not to suggest that the ADA is an efficient way to intervene in labor markets; its focus on early intervention, however, makes a lot of sense.

According to the Mashaw chapter, our experience with the Rehabilitation Act of 1973 shows little likelihood of litigation over who is disabled under the ADA. While this may or may not be a reasonable inference, it would not imply that enforcement of the ADA will have no cost to society. It would suggest, rather, that one side recognizes it cannot win and so swallows the cost. How much the ADA will eventually cost firms and the consumers of their products is an important question. In addition there are inefficiency costs to society, to the extent there are other, more cost-effective ways of achieving the goal of work for the disabled.

Much can be learned from the German experience about the advantages and disadvantages of attempting to bolster the employment of the disabled through a highly regulatory approach as contained in the ADA.[1] The Germans have a straightforward quota system. All employers with sixteen or more workers are required to hire one handicapped person for every sixteen workers. This is referred to as the 6 percent rule. Handicapped workers are defined as individuals with mental or physical impairments that reduce their capacity to work by at leat 50 percent; individuals who are at least 30 percent impaired and unemployed are also considered handicapped. Obviously this is a much broader definition of disability than the one used in our disability programs.

To lay off a handicapped worker in Germany, a special process must be followed. A firm must first notify the local labor administrator. A meeting is then held between union representatives, handicapped representatives, and management, and there must be unanimous agreement that the handicapped worker should be laid off. In addition the Veterans Bureau must concur with the decision. While most requests by employers to terminate the employment of a handicapped worker are eventually approved, the process is costly and time-consuming. (Workers who are laid off can turn to rehabilitation programs and to benefit programs which replace about three-quarters of prior earnings.)

The obligation to hire the handicapped is not ironclad. Firms can escape the requirement by paying a fine. For every quota slot not filled with a handicapped worker the firm must pay the government 150 deutsche marks (about $80 in January 1990) per month. As might be expected, there is less than full compliance with the quota.

In the mid-1980s German employers were asked why they were paying the fine rather than hiring handicapped workers, and 40 percent replied that the dismissal conditions were so stringent they preferred not to involve themselves in the process in the first place.[2] The quota system has raised the cost of firing handicapped workers, and employers are thinking twice before hiring them.

How successful has this system been in keeping disabled workers in the marketplace in Germany? The 6 percent rule worked fairly well in the 1970s, when unemployment was relatively low. In 1985, however, when unemployment was 9.3 percent (a post–World War II high for Germany), the percentage of legally handicapped workers in the economy fell to 5 percent. Hence the absolute number of handicapped workers in the labor force declined because of both the decline in total employment and the greater willingness of employers to pay fines. A policy goal of keeping handicapped workers in the labor force is more difficult—and costly—to achieve during recessions than during booms.

Ironically, as the United States embarks on further regulation of labor markets that would move in the direction of German policy, Germany is rethinking its own policies. In 1986 concern about the cost of laying off workers led to some easing of the quota system. Under the new rules a part-time handicapped worker counts as a full-time handicapped worker. In addition, handicapped apprentices—young people coming into the labor force—are counted twice. A trial work period has also been authorized for new employees, during which the special dismissal conditions for the disabled are suspended. Finally, the German government now pays the salaries of handicapped workers during their trial work period, along with any cost of job accommodation.

Note that under the ADA all accommodation costs are the responsibility of the firm.[3] Ironically, the Senate voted down amendments that would have expanded eligibility for tax credits to firms that accommodated workers protected by the ADA, because such tax credits were considered too costly to the government. Obviously this concern did not extend to private employers.[4]

Germany is committed to keeping people with severe health conditions in the workplace even if it is not efficient to do so on the basis of strictly economic criteria. Rather than obscure these costs through an off-budget, mandated program like the ADA, Germany devotes tax revenues to job accommodation. In the debate surrounding the ADA in the United States there was little serious discussion of the real costs associated with accommodation.

The ADA will reduce the flexibility of employer and employee contracts. It will also increase the cost of dismissing certain kinds of workers and hence reduce the chance for members of these protected groups to be hired in the first place.

If this argument sounds familiar, it should. It resembles the argument that was made against establishing the Equal Employment Opportunity Commission and against the EEOC's efforts to end discrimination based on race and sex. To some degree the EEOC has caused employers to be more cautious about who they hire, because they recognize that they suffer increased litigation if they fire members of protected groups. That does not mean we should not have adopted the Civil Rights Act of 1964. On the whole it was a good idea. But does it make sense to use this kind of policy to protect the disabled? I believe the negative consequences on employment will be greater because the cost of accommodating disabled workers is greater than that of accommodating people of different color or sex. There is clearly a trade-off between reducing discrimination and raising the cost of hiring disabled people. Yet in the debate over ADA almost no evidence was presented measuring the importance of discrimination in preventing handicapped workers from continuing in the labor force.

Unfortunately there is virtually no empirical research indicating to what extent employers now accommodate handicapped workers. We can, however, glean some interesting information from the Social Security Administration's 1978 survey of disability and work.[5] It is the most recent large-scale data set that includes a question on accommodation. Burkhauser and Kim studied a sample of male workers in the survey who had health conditions that limited their work activities.[6] These workers were asked whether their employers did anything to accommodate them at the onset of their work impairment. About 30 percent said yes. What effect did that have? Holding other factors constant, workers who received help continued in their jobs between two and three years longer than those who received no help, and their rate of application for disability benefits was significantly lower. In short, accommodation takes place in unregulated markets and it has positive effects on the labor market outcomes of the disabled.

No one doubts that it is better for handicapped workers to be on the job rather than on DI. But it is far from obvious that the ADA is the best way to achieve this social goal. Passage of the ADA was achieved with little or no serious discussion of its costs to society. Alternative means of achieving its ends, such as the extension of tax credits to firms that accommodate the handicapped, were ignored. It is unfortunate but probable that the blunt stick of court-enforced compliance will achieve much less than its adherents believe and will do so at significant cost to all of us.

Disability and Self-Sufficiency

A Commentary by Martin H. Gerry

The chapters by Donald O. Parsons and Jerry L. Mashaw are excellent essays on a difficult and important public policy issue—the problem of defining and determining disability. Parsons focuses on the social security disability insurance (DI) program from an economic perspective, whereas Mashaw focuses on the Americans with Disabilities Act (ADA) from a legal perspective. This commentary will attempt to provide a bridge between the two.

It is helpful to consider the definition of disability within the context of American social policy, as opposed to considering it from a purely legal or economic perspective. President Bush has laid out three basic goals for disabled individuals: to maximize economic self-sufficiency, to maximize social integration, and to maximize personal autonomy and independence. The DI program and the ADA statute have important impacts on these goals, both positive and negative.

In a sense DI's existence is an acknowledgment that we have failed to achieve these important social goals: we have not enabled many individuals with disabilities to maximize economic self-sufficiency. These people have instead entered a long-term dependency system that minimizes economic self-sufficiency and prevents social integration. Integration of disabled adults occurs primarily in the workplace. Failure to participate in work, which is a prerequisite for DI eligibility, often increases social isolation and limits personal autonomy. The ADA, by contrast, is explicitly designed to promote these three social goals.

In defining disability the federal statutes reflect three basic approaches, which are often combined and confused. According to the medical approach, which underlies the social security act programs for persons with disabilities, DI and supplemental security income (SSI), the key word is "impairment." An individual must have a condition that is clinically identifiable in a concrete, tangible way. Even concepts like mental illness and mental retardation are treated as though they were medical conditions.

89

The second approach can be described as the behavioral approach, and the key word underlying its programs and policies is "employability." The rehabilitation act is the classic example of this approach. The focus is on whether someone can work, and this determines who gets services and who does not. The focus is not on disability in the medical sense.

Finally, in what I would describe as the sociological approach, the key word is discrimination; the focus here is on how people are treated. Individuals are disabled if they have had their rights limited or have been precluded from doing something meaningful. Many people who are impaired in a medical sense may never have been disabled in a sociological sense and may never have experienced discrimination. That is, they have not been excluded from employment and may not have been affected adversely by being considered disabled.

There is also a distinction between the concepts of disability and handicap. The World Health Organization uses the term "handicap" to mean a limitation imposed on an individual by the environment in which he operates or seeks to operate. A person can be handicapped in one environment and not in another, while in a medical sense being equally disabled. The term "handicap" thus focuses on the operational relationship between the environment and the individual. The term "disability," by contrast, is used by the World Health Organization to refer to some inherent characteristic of the individual. The World Health Organization, then, uses something close to a medical definition for disability and a sociological definition for handicap.

Regardless of how one defines disability, however, the condition is not static; it is affected in important ways by the passage of time. I realized this most clearly while serving with Carolyn Weaver on the Disability Advisory Council of the Department of Health and Human Services in 1987. Most of the research available to us—and that was not much—suggested that the longer an individual is out of the work force, the less likely it is that he will ever return to the work force. There appears to be a direct correlation between the passage of time and the individual's perception of his or her ability to work. An expert panel of physicians told the council that the medical facts about an individual's condition accounted for considerably less than 50 percent of the variability in work outcomes. The more important factor, in their view, was attitude.

In considering the incentive effects of provisions in the law, such as the waiting period for DI benefits, economists should not overlook the effects of time on disability. A waiting period may discourage some people from dropping out of the work force to get benefits, but for those who do drop out the time lost meeting the waiting period may undermine subsequent efforts to return to work. The time factor also needs to be kept in mind in evaluating the costs and benefits of the ADA. If successful the ADA will allow people who become disabled to remain at work and to avoid some of the potentially deleterious effects of time on the severity of disability.

A high percentage of people who now leave the work force and go on DI benefit rolls could probably be retained in the work force with job accommodation. Many people choose to drop out for reasons having more to do with job dissatisfaction than with their underlying physical or mental impairments. The ADA, like section 503 and section 504 of the rehabilitation act, strongly encourages employers to make job modifications as peoples' disabilities occur. We should look at strategies to reward employers for doing precisely that. In an active and affirmative way we should couple the DI program with the ADA, using the ADA statute to help keep people in the work force and thus reduce the number who apply for DI benefits.

Even for an unchanging definition of disability, such as the one used in the DI program, the composition of people served by a program for the disabled can change dramatically over the years. Traditionally DI was a program primarily for older people who had been in the work force for many years and who would probably collect benefits for a short period. The program's risk exposure averaged about eight years for these people. (The average age of a beneficiary just a few years ago was fifty-seven, and DI beneficiaries convert to the old-age and survivors insurance program to receive benefits beginning at age sixty-five.)

In the last five years, however, there has been a remarkable change in the nature of the DI program: approximately one-third of the new beneficiaries awarded DI benefits have been under age forty-five, and almost a quarter of them have had mental impairments. These people may be on benefits for as long as forty years. This suggests a major change in the economic impact of DI on the government's costs and revenues.

For the past year and a half I have been conducting research on what could loosely be described as the cost of failure, or the cost of keeping people dependent. This research has focused on the social security programs, DI and SSI. Let us consider those people under age thirty-five who are now on the SSI or DI rolls. Assuming the current rate at which beneficiaries leave the benefit rolls to return to work (less than 0.5 percent per year), the cost of state and federal benefits (including Medicaid, Medicare, and food stamps) over these beneficiaries' lifetimes will be slightly over $1 trillion in current dollars. That is to say, it costs a lot of money to dribble out sub-poverty-level benefits to a substantial number of people for a long period of time. Were we a private insurer, we would be trying to control that risk exposure and reduce the likelihood of our having to spend all that money.

Let us now turn to the chapter by Richard Burkhauser, in which he draws parallels to disability policy in the Federal Republic of Germany. First the German system has so many problems and German disability policy is so inherently flawed that it would be unwise to draw conclusions for U.S. policy. It simply is not clear whether these problems stem from the high cost of firing disabled people or from the fact that quota hiring systems are not very good ways to expand the employment of the disabled. Consequently it is less clear what conclusions can be drawn for the ADA, which does not involve any quotas.

Although Burkhauser raises a number of concerns about the ADA, the act's employment provisions are fashioned after sections 503 and 504 of the Rehabilitation Act of 1973, which already apply to federal contractors and recipients of federal funds. As Mashaw has noted, there has been very little case law under these sections of the rehabilitation act, which suggests that there was a serious problem and that employers have been responding positively to the law. The implementation of the employment provisions in sections 503 and 504 was arguably more successful and less controversial than any other aspect of the rehabilitation act.

Finally, let us consider the costs of accommodation, which, for as long as I have been in Washington, have been described as onerous. People seem to forget that private employers already spend a substantial amount of money in accommodating nondisabled employees. There has been a rapid expansion of corporate-financed programs dealing with alcohol abuse, drug treatment, child care, and other personal problems. Larger corporations in the United States, as in Japan, are increasingly looking at employees in a more holistic way. Installing a ramp in a building is not necessarily more expensive than many other measures that corporations are already committed to in serving a variety of employee needs.

The prospects for true reform of our disability policy in the United States do not look encouraging. At present we have a huge dependency industry that has been created by the long-term unemployment of a lot of disabled people. This poses a real obstacle to reform. In the case of community-based Medicaid reform, for example, the idea of providing treatment outside institutions creates concern among prominent congressmen about the impact on employ-ment of the people running the institutions. Increasingly, the implicit question behind policy debates is, What are we going to do with all these social workers? What will happen to all these people who are in the dependency business if a large number of disabled people go back to work and no longer need them? The ADA may be a step toward overcoming this obstacle.

Notes

PREFACE, *Carolyn L. Weaver*

1. Americans with Disabilities Act of 1990, 42 U.S. Code 12101 (July 26, 1990).

2. Testimony of Ronald A. Lindsey before the Republican Study Committee, U.S. House of Representatives, February 7, 1990.

The absence of a single, broadly accepted measure of the disabled population that can be quantified, verified, and used for careful empirical analysis has hampered research efforts as well. No careful empirical study of the nature and extent of labor market discrimination against the disabled currently exists, for example.

CHAPTER 1: INCENTIVES VERSUS CONTROLS, *Carolyn L. Weaver*

1. The federal government spends roughly $75 billion annually on an array of programs for the disabled. (Estimate based on 1986 figure of $60 billion, adjusted for average budget growth between 1986 and 1990.) See National Council on the Handicapped, *On the Threshold of Independence*, a report to the president and the Congress (Washington, D.C.: Government Printing Office, January 1988), p. 2, and President of the United States, *Economic Report of the President* (Washington, D.C.: Government Printing Office, January 1990), p. 384.

2. In 1989, approximately $30 billion was spent on DI and SSI (disabled workers), and roughly an equal amount was spent on Medicare and Medicaid for the disabled. See Board of Trustees of the Federal Old-Age and Survivors Insurance and Disability Insurance Trust Funds, *1989 Annual Report* (Washington, D.C.: Government Printing Office, 1989); U.S. Department of Health and Human Services, Social Security Administration, *Social Security Bulletin: Annual Statistical Supplement, 1988* (Washington, D.C.: Government Printing Office, 1989), p. 313; and U.S. Department of Health and Human Services, Health Care Financing Administration, Office of the Actuary.

3. For a summary of these provisions, see "Social Security Disability Amendments of 1980: Legislative History and Summary of Provisions," *Social Security Bulletin*, no. 44 (April 1980), pp. 1–18. See also U.S. Congress, House, *Conference Report on the Disability Amendments of 1980*, 96th Congress, 2d session, H. Rept. 96–944.

Work incentives were among the key issues studied by the Disability Advisory Council, appointed by the secretary of health and human services in 1986. See *Report of the Disability Advisory Council* (Washington, D.C.: Social Security Administration, March 1988). See also S. 1358, introduced on July 19, 1989, which is designed to expand work incentives in the DI program.

4. Presently, the targeted jobs tax credit is 40 percent of the first year's wages up to $6,000, for a maximum credit of $2,400. Section 190 allows a deduction for up to $35,000 annually. The EITC provides a refundable tax credit of 14 percent against the first $6,500 of earned income, for a maximum credit of $910; the credit is reduced for incomes above $10,240, and eliminated for incomes exceeding $19,340.

Another provision in the law, adopted in 1966, permits the payment of a special minimum wage to disabled workers. With approval from the Department of Labor, employers can offer 50 percent or more of the federal minimum wage to such workers. This was adopted in recognition of the fact that the minimum wage could curtail work opportunities for the disabled. See 24 C.F.R. V sec. 524, July 1, 1988. At present, approximately 300,000 individuals are being paid in accordance with this provision. Figure provided by the Hour and Wage Division of the U.S. Department of Labor.

5. U.S. Department of Health and Human Services, Social Security Administration, *Social Security Bulletin: Annual Statistical Supplement, 1980* (Washington, D.C.: Government Printing Office, 1981), and U.S. Congress, Senate Finance Committee, *Staff Data and Materials Related to the Social Security Disability Insurance Program*, 97th Congress, 2d session, 1982. Committee Print 97–16.

6. This figure also includes those who left the rolls because of medical recovery. See Ralph Treitel, "Recovery of Disabled Beneficiaries: A 1975 Followup Study of 1972 Allowances," *Social Security Bulletin*, no. 42 (April 1979), pp. 3–23.

The relationship between the level of DI benefits and the rate of male nonparticipation in the labor force has been quantified empirically by Donald O. Parsons in "The Decline of Male Labor Force Participation," *Journal of Political Economy*, vol. 88, no. 1 (February 1980), pp. 117–34. See also Jonathan S. Leonard's review of his and other work on the same issue in "Labor Supply Incentives and Disincentives for Disabled Persons," in Monroe Berkowitz and M. Anne Hill, eds., *Disability and the Labor Market: Economic Problems, Policies, and Programs* (Ithaca, N.Y.: ILR Press, Cornell University, 1986).

7. The Americans with Disabilities Act of 1990, 42 U.S. code 12101 (July 26, 1990).

8. For a statement of this view, see, for example, remarks by Senator Edward Kennedy, *Congressional Record*, 101st Congress, 1st session, September 7, 1989: S10717–18: "Our society is still infected by the ancient, now almost subconscious assumption that people with disabilities are less than fully human and therefore not fully eligible for the opportunities, services, and support systems, which are available to others as a matter of right. The result is massive, societywide discrimination." See also U.S. Commission on Civil Rights, *Accommodating the Spectrum of Individual Abilities*, Clearinghouse Publication, no. 81 (Washington, D.C.: CCR, September 1983).

9. In competitive markets, individuals are treated equally (and the outcomes are efficient) when the marginal cost of each worker equals the marginal benefit; with fixed costs, such as those associated with certain types

of accommodations, the worker's total product must equal or exceed his total cost.

10. See Gary S. Becker, *The Economics of Discrimination*, 2d ed. (Chicago: University of Chicago Press, 1971), for a careful analysis of the economic theory of discrimination. See also William M. Landis, "The Economics of Fair Employment Laws," *Journal of Political Economy*, vol. 76 (July/August 1968), pp. 507–45.

11. Inefficiencies would stem from compliance costs, such as the costs of modifying screening, application, and hiring procedures, or from administrative or judicial actions that made it costly for employers to defend themselves against erroneous claims of discrimination, thus increasing the cost of laying off less productive workers.

12. Americans with Disabilities Act, sec. 102(a).

13. Americans with Disabilities Act, sec. 101(8).

14. Title 7 remedies include injunctive relief and reinstatement or hiring, with or without back pay (up to two years). It does not generally provide for compensatory or punitive damages or jury trials, as provided under section 1981 of the civil rights act. However, legislation now pending in Congress, the Civil Rights Act of 1990, would increase monetary damages and expand the availability of jury trials.

Following title 7, the employment provisions of the ADA will provide for administrative enforcement by the EEOC (which is to attempt voluntary conciliation before bringing a civil action), and also provide a private right of action in cases where the EEOC dismisses the charge, or fails to reach a conciliation or to file a suit within 180 days.

For more on the provisions of the Civil Rights Act of 1964 and the ADA, see Charles V. Dale, "Remedies and Standing to Sue under S. 933, the 'Americans with Disabilities Act of 1989,' " U.S. Congressional Research Service Report for Congress (Washington, D.C.: CRS, May 26, 1989), and Nancy Lee Jones, "The Americans with Disabilities Act (ADA): An Overview of Selected Major Legal Issues," U.S. Congressional Research Service, report for Congress (Washington, D.C.: CRS, July 25, 1989).

15. The Rehabilitation Act of 1973, Public Law No. 93-112, 87 Stat. 355, and 5 C.F.R. sec. 84, October 1, 1988.

16. Americans with Disabilities Act, sec. 102(b)(5)(A).

17. Americans with Disabilities Act, sec. 101(9).

18. Americans with Disabilities Act, sec. 101(10).

19. Section 504 applies to any private (or state or local) employer receiving funds. See 5 C.F.R. sec. 84, appendix A, October 1, 1988.

20. Accommodation can also take place indirectly, as when a firm streamlines and simplifies some component of its production process for purposes of quality control. In so doing it may well create job opportunities for a new pool of workers, the mentally retarded. Alternatively, when a firm computerizes its operations to enhance the speed, accuracy, and breadth of its information-distribution system, it may well have created job opportunities for a new pool of workers, the blind or the mobility-impaired. In a real sense, these workers have been "accommodated" at no cost to themselves.

21. "Positive returns" is used to connote a return that is at least as good as could be earned on alternative investments; employees and employers are made better off in an ex ante sense.

22. The minimum wage limits the adjustments that can be made in the wage component of the compensation package, the employment opportunities of unskilled workers with disabilities, and the likelihood of accommodation.

23. The term "reasonable accommodation" originated in 1977, with the issuance of the regulations implementing section 504 of the rehabilitation act. For more on the case law, see *Accommodating the Spectrum of Individual Abilities*, pp. 102, 104; and Robert L. Burgdorf, Jr., ed., *The Legal Rights of Handicapped Persons: Cases, Materials, and Text* (Baltimore: Paul H. Brookes, Publishers, 1980).

24. Americans with Disabilities Act, sec. 102(3).

25. Americans with Disabilities Act, sec. 3(2).

26. "Major life activities means functions such as caring for one's self, performing manual tasks, walking, seeing, hearing, speaking, breathing, learning, and working," 45 C.F.R. subtitle A, sec. 84.3, October 1, 1988.

See statements by Senators David Pryor and William Armstrong on the breadth and uncertainty of the definition of disability. *Congressional Record*, 101st Congress, 1st session, September 7, 1989: S10741, S10753–55, and S10785–56.

27. It is clear only in the case of food-handling positions. There was a spirited debate in the Senate on the question of whether the ADA protected individuals with AIDS who wished to work in restaurants in positions involving food handling. A provision was added to the legislation clarifying that firms may refuse to assign to food handling positions people with contagious or infectious diseases that are transmitted through food handling; the secretary of health and human services, however, shall make the determination of which diseases fall into this category. (See section 103(b) and 103(d) of the Americans with Disabilities Act.) It is most unlikely (at this time) that the secretary would find AIDS to be such a disease. More generally, the ADA (section 103(b)) would allow firms to utilize qualification standards that disqualify from employment people with currently contagious diseases or infections that pose a "direct threat" to the health or safety of other individuals in the work place. In keeping with a recent Supreme Court decision interpreting section 504 of the rehabilitation act, the employer would have the obligation to make reasonable accommodations to the individual's condition (and retain the individual) if such accommodations could remove this threat.

In *School Board of Nassau County v. Arline* 107 S.Ct. 1123 (1987), the Supreme Court found that a person with tuberculosis may be protected under section 504 of the rehabilitation act. The Department of Justice responded to this decision by issuing a memorandum regarding the coverage of people with AIDS or the HIV. The memo stated that such individuals are covered if they are able to perform the duties of the job and do not constitute a direct threat to the health or safety of others. Employers have the duty to make reasonable accommodations, and if such accommodations cannot remove the threat to the health or safety of others then such individuals are not protected under the law. See Mary F. Smith, "Section 504 of the Rehabilitation Act: Statutory Provisions, Legislative History, and Regulatory Requirements," U.S. Congressional Research Service report for Congress (Washington, D.C.: CRS, January 19, 1989).

28. The bill does not specify responsibilities for disabled workers such as following prescribed medical treatments to control or reduce symptoms or utilizing rehabilitation services. In addition, disabled workers have no obligation to remain with a firm after costly accommodations have been made.

29. Since the law is premised on the assumption of widespread discrimination and it includes, in the definition of discrimination, the failure to accommodate reasonably yet does not allow for a balancing of costs and benefits, I conclude that the legislation is likely to affect the wages and accommodations offered by firms that are not economic discriminators as well as firms that are—inducing too much accommodation (and wages) by nondiscriminators and increasing the wages and accommodation by discriminators.

The unskilled, severely impaired will be particularly expensive to hire because of the constraint imposed by the minimum wage.

30. A careful study of the net impact of the legislation on the employment and wages of disabled people in private firms would also have to consider the effects of the rehabilitation act, which already applies to firms receiving federal funds, as well as the various state antidiscrimination laws, the provisions of which are not uniform.

31. In Germany, where a quota system applied to disabled workers in the early 1980s, employers found it difficult to lay off unproductive disabled workers and thus were discouraged from hiring the disabled. A 1986 amendment deregulated the relation between disabled employees and employers in German labor markets. See Petri Hirvonen and Richard V. Burkhauser, "German Disability Policy in a Time of Economic Crisis," Vanderbilt University Working Paper (May 1989).

32. Proponents of the ADA cite figures from a Department of Labor study on the cost of complying with section 503 of the rehabilitation act (which applies to federal contractors) to suggest the relatively low expense of the reasonable accommodation requirement. The study reported that in a sample of firms, in half the cases of accommodation, the cost was zero; another 30 percent of the cases involved costs under $500. Fewer than 5 percent involved costs over $5,000. (The most expensive accommodations were made for the blind and those in wheelchairs.) In reporting accommodation costs, however, firms tended to report only out-of-pocket costs—excluding management time, the time of engineers and employees, and even, in some cases, the cost of materials for construction. Furthermore, the disabled people already hired are likely to be less costly to accommodate than those not yet hired. No effort was made to quantify the benefits of the provision or the extent to which these costs would have been incurred absent the law. See Berkeley Planning Associates, *A Study of Accommodations Provided to Handicapped Employees by*

Federal Contractors: Final Report, 2 vols. Prepared by under contract no. J-9-E-1-0009 (1982); and Frederick Collignon, "The Role of Reasonable Accommodation in Employing Disabled Persons in Private Industry," in *Disability and the Labor Market: Economic Problems, Policies, and Programs,* Monroe Berkowitz and M. Anne Hill, eds. (Ithaca, N.Y.: ILR Press, Cornell University, 1986), pp. 196–241.

33. For more on this, see Carolyn L. Weaver, "Disability Policy in the 1980s and Beyond," in *Disability and the Labor Market: Economic Problems, Policies, and Programs,* Monroe Berkowitz and M. Anne Hill, eds. (Ithaca, N.Y.: ILR Press, Cornell University, 1986), pp. 29–63.

34. For the seminal article on the role of decentralized markets in solving these information problems, see F. A. Hayek, "The Use of Knowledge in Society," *American Economic Review,* vol. 35, no. 4 (September 1945), pp. 519–30.

35. This is true in the case of reforming vocational rehabilitation, no less than in the case of reforming the tax code as it affects employers and the working disabled, or reforming the various income support programs for the nonworking disabled. The introduction of vouchers in the rehabilitation program, for example, would introduce freedom of choice for users and competition in supply among providers, which could only enhance the success of these programs. This proposal is in keeping with the recommendation of the Disability Advisory Council: "SSA should promote active competition among public and private [rehabilitation] agencies." See its *Report of the Disability Advisory Council.*

Chapter 2: Accommodation and the Labor Market,
Sherwin Rosen

1. Americans with Disabilities Act of 1990, 42. U.S. Code 12101 (July 26, 1990).

2. See Donald O. Parsons, "The Decline in Male Labor Force Participation," *Journal of Political Economy,* vol. 88, no. 1 (February 1980), pp. 117–34; Richard V. Burkhauser and Robert H. Haveman, *Disability and Work: The Economics of American Policy* (Baltimore, Md.: Johns Hopkins University Press, 1982); John Bound and Tim Waldmann, "Disability Transfers and the Labor Force Attachment of Older Men: Evidence from the Historical Record," working paper (Ann Arbor: University of Michigan, 1989); and Donald O. Parsons, "The Male Labor Force Participation Decision: Health, Reported Health and Economic Incentives," *Economica,* vol. 49 (February 1982), pp. 81–91.

3. See Carolyn L. Weaver, "Disability Policy in the 1980s and Beyond," in Monroe Berkowitz and M. Anne Hill, eds., *Disability and the Labor Force Market: Economic Problems, Policies, and Programs* (Ithaca, N.Y.: ILR Press, Cornell University, 1986), pp. 29–63.

4. See John Bound, "The Health and Earnings of Rejected Insurance Disability Applicants," *American Economic Review,* vol. 89, no. 3 (June 1989), pp. 482–503; and Robert H. Haveman and Barbara L. Wolfe, "Disability

Transfers and Early Retirement: A Causal Relationship," *Journal of Public Economics* (June 1984), pp. 47–66.

5. See Robert F. Cotterman and John Raisian, "The Incidence of Disability, 1970 to 2020: A Public Policy Dilemma?" A report submitted to the National Council on the Handicapped (Washington, D.C.: Unicon Research Corporation, May 1988).

6. See Parsons, "Decline in Male Labor Force Participation," pp. 117–34.

7. See Cotterman and and Raisian, "Incidence of Disability."

8. See Marvin H. Kosters, "Wages and Demographics," to appear in *Workers and Their Wages: Changing Patterns in the United States* (Washington, D.C.: American Enterprise Institute, forthcoming).

9. See Frederick C. Collignon, "The Role of Reasonable Accommodation in Employing Disabled Persons in Private Industry," in Monroe Berkowitz and M. Anne Hill, eds., *Disability and the Labor Force Market: Economic Problems, Policies, and Programs* (Ithaca, N.Y.: ILR Press, Cornell University, 1986), pp. 196–241.

CHAPTER 3: WORKFARE-WELFARE DILEMMA, *Walter Y. Oi*

1. The concept of health capital was introduced by Michael Grossman, in "The Concept of Health Capital and the Demand for Health," *Journal of Political Economy*, vol. 80 (March 1972), pp. 223–55.

2. On the importance of functional-activity limitations, medically determinable impairments, and the setting in which the individual operates see, among others, L. Haber, "Identifying the Disabled: Concepts and Methods in the Measurement of Disability," *Social Security Bulletin*, no. 30 (December 1967), pp. 17–35 and Saad Z. Nagi, *Disability and Rehabilitation: Legal, Clinical, and Self Concepts and Measurement* (Columbus: Ohio State University Press, 1969).

3. Americans with Disabilities Act of 1990, 42 U.S. Code 12101, sec. 3(2) (July 26, 1990).

4. Section 223b(1), Social Security Act as amended, 42 U.S.C. 5433 (b).

5. The SGA limit is higher for the blind.

6. The empirical studies can be found in: Jonathan S. Leonard, "The Social Security Disability Insurance Program and Labor Force Participation," National Bureau of Economic Research Working Paper, no. 392 (Cambridge, Mass., 1979); Leonard, "Disability Incentives and Disincentives for the Disabled," in Monroe Berkowitz and M. Anne Hill, eds., *Disability and the Labor Market: Economic Problems, Policies, and Programs* (Ithaca, N.Y.: ILR Press, Cornell University, 1986), pp. 64–96; and Donald O. Parsons, "The Decline in the Male Labor Force Participation," *Journal of Political Economy*, vol. 88 (February 1980), pp. 117–34.

Data on the size and composition of the DI and SSI populations can be found in John L. McCoy and Kerry Weems, "Disabled Worker Beneficiaries and Disabled SSI Recipients: A Profile of Demographic and Program Characteristics," *Social Security Bulletin*, vol. 52, no. 5 (May 1989), pp. 16–28.

7. The incidence of severe disabilities is surely subjective. Richard Burk-

hauser calculated the ratio of disability transfer recipients (who were judged to be incapable of work) to the labor force in three countries. In the United States it climbed from .027 in 1970 to .039 in 1982. In Sweden it was .049 in 1970 and .068 in 1980. Finally, in the Netherlands the ratio of disability transfer recipients to the labor force was .087 in 1974 and .130 in 1980. See Richard V. Burkhauser, "Disability Policy in the United States, Sweden, and the Netherlands," in Monroe Berkowitz and M. Anne Hill, eds., *Disability and the Labor Market: Economic Problems, Policies, and Programs* (Ithaca, N.Y.: ILR Press, Cornell University, 1986), pp. 262–84. Variations of these magnitudes cannot be explained by differences in the incidence of medical impairments and functional-activity limitations over time or across countries.

8. See Robert L. Bennefield and John M. McNeil, "Labor Force Status and Other Characteristics of Persons with a Work Disability: 1981 to 1988," U.S. Bureau of the Census, series P-23, no. 160 (July 1989). In this report, an individual is said to have a work disability if one or more of the following conditions is met. The individual (1) has a condition that prevents him from working or limits the kind or amount of work he can do; (2) ever retired or left a job for health reasons; (3) did not work in the survey week because of a long-term illness or disability that prevented the performance of any kind of work; (4) did not work in the previous year because ill or disabled; (5) is under sixty-five years of age and covered by Medicare; and (6) is under sixty-five and a recipient of SSI. If one or more of the last four criteria are met, the individual was classified as having a severe disability. The study was confined to the working-age population of persons sixteen to sixty-four years of age.

Bennefield and McNeil pointed out that surveys designed to study disability or health status typically produce higher incidence rates. The percentage with a work disability was 8.6 percent in the 1988 Current Population Survey, 12.1 percent in the 1984 survey of the income and program participation, and 11.5 percent in the 1983–1985 health interview survey.

9. G. Dejong, A. I. Batavia, and R. Griss, "America's Neglected Health Minority: Working-Age Persons with Disabilities," *Milbank Quarterly*, vol. 67, no. 2, supplement (1989), pp. 311–51.

10. The advances in medical science that lowered mortality rates raised the proportion of persons who survived with functional-activity limitations. This relation was established by L. Verbrugge, "Longer Life but Worsening Health: Trends in Health and Mortality of Middle-Aged and Older Persons," *Milbank Memorial Fund Quarterly*, vol. 62, no. 3 (1984), pp. 475–519.

11. Louis Harris and Associates, Inc., *The ICD Survey of Disabled Americans: Bringing Disabled Americans into the Mainstream* (New York: International Center for the Disabled, March 1986).

12. Walter Y. Oi, "Three Paths from Disability to Poverty," Technical Analysis Paper, no. 58, U.S. Department of Labor, Office of the Assistant Secretary for Policy Evaluation and Research (Washington, D.C., October 1978).

13. Paula Franklin, "Impact of Disability on the Family Structure," *Social Security Bulletin*, vol. 40, no. 5 (May 1977), pp. 3–18.

14. Robert H. Haveman and Barbara Wolfe, "The Economic Well-Being of the Disabled, 1962 to 1984," *Journal of Human Resources*, vol. 25 (Winter 1990).

15. In principle, K could be described by a vector identifying all of the pertinent organisms and functions. For expository ease, I assume that K is described by a single index. For more on the concept of health capital, see Grossman, "Concept of Health Capital."

16. If K is described by a vector, a destruction of some elements of K need not affect an individual's market productivity. For example, the loss of hearing will have effects on the productivity and wages of a truck driver different from its effects on a nurse.

17. The time dimension could be included in a definition of work capacity. Individuals with smaller stocks of discretionary time or greater uncertainty in "well time" on any given day could be directed to special training programs and counseled to follow vocations where the pay penalties for flextime and short work schedules are smaller. They could also be granted more tax credits to procure aids that economize on the use of time.

18. Frank Bowe, "Out of the Job Market: A National Crisis" (Washington, D.C.: President's Committee on Employment of the Handicapped, 1986).

19. Collignon cites a 1948 Department of Labor study in which the performance of 11,000 impaired workers was compared to that of a control group of 18,000 able-bodied workers in matched jobs. The impaired workers were slightly more productive (1 percent above the nondisabled workers), had a slightly higher absenteeism rate (3.8 versus 3.4 absences per 100 working days), and had higher quitting rates (3.6 versus 2.6 quits per 100 employees). However, the impaired workers constituted a self-selected sample representing only a minority of all disabled persons. The difficulty of identifying the target population of disabled persons probably accounts for the absence of a reliable statistical study. For details see Frederick C. Collignon, "The Role of Reasonable Accommodations in Employing Disabled Persons in Private Industry," in Monroe Berkowitz and M. Anne Hill, eds., *Disability and the Labor Market* (Ithaca, N.Y.: ILR Press, Cornell University, 1986), pp. 196–241.

20. Americans with Disabilities Act of 1990, sec. 102(5)(A) and (5)(B).

21. Collignon, "Role of Reasonable Accommodations," p. 199.

22. U.S. Department of Health, Education, and Welfare, Rehabilitation Services Administration, "The Comprehensive Needs Study" (Washington, D.C.: Urban Institute, 1975), cited by Collignon, "Role of Reasonable Accommodations," p. 16.

23. Collignon, "Role of Reasonable Accommodations."

24. Ibid., tables 8.4 and 8.5. The percentage of disabled workers in a firm's work force is higher for larger employers who are likely to spend more for workplace accommodations.

25. Collignon, "Role of Reasonable Accommodations," p. 205.

26. Americans with Disabilities Act of 1990, sec. 302(b)(1)(A).

27. If the lift is located at the back of the bus, each pick-up or drop takes around four to five minutes. The externalities include not only the added time costs incurred by riders, but also the cost of disrupting the bus schedule. A fuller discussion of the transbus case can be found in Richard V. Burkhau-

ser and Robert H. Haveman, *Disability and Work: The Economics of American Policy* (Baltimore, Md.: Johns Hopkins University Press, 1982), pp. 82–85.

28. Some who were denied readmission to DI appealed their adverse decisions. Many were later readmitted to DI. The experience was evidently unpleasant for all parties. It would seem preferable to reduce such conflicts in the future, possibly by drawing a distinction between permanent and temporary disabilities in a manner analogous to the state worker compensation laws.

29. Damages under the ADA are the same as those under sections 705–710 of the Civil Rights Act of 1964, which would be modified by the Civil Rights Act of 1990, now pending in Congress.

30. Bye and Riley report that 12.3 percent of new entrants to the DI rolls from 1972 to 1981 died during the two-year waiting period before receipt of Medicare benefits. B. V. Bye and G. F. Riley, "Eliminating the Medicare Waiting Period for Social Security Disabled Worker Beneficiaries," *Social Security Bulletin*, no. 52 (May 1989), pp. 2–15.

31. To promote a return to work and perhaps to avoid confrontations in disability reviews, the SGA limit has recently been raised to $500 a month, and people who return to work have been permitted to keep their Medicare coverage for up to forty-eight months. Apparently the government is prepared to uncouple cash and Medicare benefits for those already on the rolls, but not for new entrants.

32. Most private insurance carriers will not allow an employer to exempt some workers from his policy because of the adverse selection risk; i.e., the workers in the best health may be exempted, leaving the carrier with the unhealthy, high-risk employees. The situation is different when the carrier is asked to exempt a disabled person who has already been awarded Medicare benefits by SSA. Excluding end-stage renal disease and AIDS, the cost of Medicare is around $2,000 a year for each disabled recipient.

33. National Council on the Handicapped, *Toward Independence*, a report to the President and to the Congress (Washington, D.C.: NCH, 1986).

34. Charles Murray, *Losing Ground: American Social Policy 1950 to 1980* (New York: Basic Books, 1984).

CHAPTER 4: POLICY AND THE RETURN TO WORK, *Jonathan S. Leonard*

1. Jonathan S. Leonard, "On the Decline in the Labor Force Participation Rates of Older Black Males" (honors thesis, Harvard University, 1976).

2. See John Bound, "The Health and Earnings of Rejected Disability Insurance Applicants," *American Economic Review*, vol. 79 (June 1989), pp. 842–503; Jonathan S. Leonard, "The Social Security Program and Labor Force Participation," National Bureau of Economic Research Working Paper, no. 392 (Cambridge, Mass., 1979); and Donald O. Parsons, "The Male Labor Force Participation Decision: Health, Reported Health, and Economic Incentives," *Economica*, vol. 49 (February 1982), pp. 81–91. See also Robert Haveman and Barbara Wolfe, "Disability Transfers and Early Retirement: A Causal Relationship," *Journal of Public Economics*, vol. 22 (June 1984), pp. 47–66; Jerry

L. Mashaw, *Bureaucratic Justice: Managing Social Security Disability Claims* (New Haven, Conn.: Yale University Press, 1983); and Carolyn L. Weaver, "Social Security Disability Policy in the 1980s and Beyond," in Monroe Berkowitz and M. Anne Hill, eds., *Disability and the Labor Market: Economic Problems, Policies, and Programs* (Ithaca, N.Y.: ILR Press, Cornell University, 1986), pp. 29–63.

3. Disability Advisory Council, *Report of the Disability Advisory Council* (Baltimore, Md.: U.S. Department of Health and Human Services, Social Security Administration, 1988), p. 67.

4. U.S. Department of Health and Human Services, Social Security Administration, *Social Security Bulletin: Annual Statistical Supplement, 1988* (Washington, D.C.: U.S. Government Printing Office, 1988), p. 280.

5. See Leonard, "The Social Security Program and Labor Force Participation," and Parsons, "Male Labor Force Participation Decision."

6. U.S. Congress, House, Committee on Ways and Means, *Background Material and Data on Programs within the Jurisdiction of the Committee on Ways and Means, 1989*, 101st Congress, 1st session, March 15, 1989. Committee Print 101–4.

7. Ibid., table 20, p. 711.

8. Roxanne M. Andrews, Martin Ruther, David K. Baugh, Penelope L. Pine, and Marilyn P. Rymer, "Medicaid Expenditures for the Disabled under a Work Incentive Program," *Health Care Financing Review*, vol. 9 (Spring 1988), pp. 1–8.

9. *Nelson v. Thornberg*, 567 F. Supp. 369 (E. D. Pa. 1983); affirmed 732 F. 2d 146 (3d Cir. 1984); cert. denied 469 U.S. 1188 (1985).

10. Remedies available under the ADA follow the Civil Rights Act of 1964, and so are tied to any subsequent amendments to that act. The failed Civil Rights Act of 1990, for example, proposed to increase access to monetary damages and jury trials, provisions that would have applied equally to the disabled.

11. Jonathan S. Leonard, "The Effectiveness of Equal Employment Law and Affirmative Action Regulation," in R. G. Ehrenberg, ed., *Research in Labor Economics*, vol. 8, part B (Greenwich, Conn.: JAI Press, 1987), pp. 319–50, and "Affirmative Action in the 1980s: With a Whimper not a Bang" (Unpublished paper, University of California, Berkeley, 1987).

12. The incentives for such differential treatment of disabled applicants and employees who become disabled are reversed under the targeted jobs tax credit program. Employers gain tax advantages for hiring disabled people receiving SSI or DI benefits but not for continuing to employ those who become disabled. The extent to which this has actually reduced the differential treatment is not known, but probably few employers have made use of the credit for the disabled.

13. Veterans Readjustment Act of 1974, 38 U.S. Code, sec. 42.

DISABILITY IN SOCIETY, Evan J. Kemp, Jr.

1. This is their estimate of the cost (in 1986) of federal programs, private individual and employer-provided programs, and indemnity programs for

the disabled. See Monroe Berkowitz, David Dean, Dale Hanks, and Stanley Portny, *Enhanced Understanding of the Economics of Disability*, Final Report to the Virginia Department of Rehabilitative Services (Richmond, Va., August 1988). See also Monroe Berkowitz and Carolyn Green, "Disability Expenditures," *Rehabilitation Review* (Spring 1989).

CHAPTER 5: IN SEARCH OF THE DISABLED, *Jerry L. Mashaw*

1. Americans with Disabilities Act of 1990, 42 U.S. Code 12101 (July 26, 1990).

2. Jerry L. Mashaw, *Bureaucratic Justice: Managing Social Security Disability Claims* (New Haven, Conn.: Yale University Press, 1983).

3. For the most recent of the many recommendations that have been made for the improvement of the SSA disability process, see the *Report of the Disability Advisory Committee to the Commissioner of Social Security* (Baltimore, Md.: U.S. Department of Health and Human Services, Social Security Administration, June 25, 1989).

4. See "Symposium, Judicial Review of Social Security Disability Decisions: A Proposal for Change," *Texas Law Review*, vol. 11, no. 215 (1980).

5. Americans with Disabilities Act of 1990, sec. 3(2).

6. Citation to the relevant statutes in forty-seven states and the District of Columbia appears in "Handicapped Workers: Who Should Bear the Burden of Proving Job Qualifications?" *Main Law Review*, vol. 38, no. 135, p. 150 n. 66.

7. See statement of Senator Harkin, 135 *Congressional Record*, 101st Congress, 1st session, May 9, 1989: S4986.

8. The regulations referred to appear at 54 *Federal Register*, January 23, 1989: 3232 and at 42 *Federal Register* May 4, 1977: 22676.

9. 480 U.S. 273 (1987).

10. 497 F. Supp. 1088 (D.C. Hawaii 1980).

11. Torres v. Bolger, 610 F. Supp. 593 (D.C. Tex. 1985), for example, rejects a claim that lefthandedness is a disability within the meaning of the Rehabilitation Act of 1973.

12. See George Priest and William Klein, "The Selection of Disputes for Trial," *Journal of Legal Studies*, vol. 13, no. 1 (1984).

13. Carolyn L. Weaver, chap. 1 of this book.

14. See, generally, John E. Calfee and Richard Craswell, "Some Effects of Uncertainty on Compliance with Legal Standards," *Virginia Law Review*, vol. 70, no. 965 (1984); Guido Calabresi and Alvin Klevorick, "Four Tests for Liability in Torts," *Journal of Legal Studies*, vol. 14, no. 585 (1985); Louis Kaplow, "Optional Government Policy toward Risk Imposed by Uncertainty Concerning Future Government Action," Harvard Program in Law and Economics Discussion Paper, no. 20 (Cambridge, Mass., August 1986).

15. See, generally, Frederick C. Collignon, "The Role of Reasonable Accommodation in Employing Disabled Persons in Private Industry," in Monroe Berkowitz and M. Anne Hill, eds., *Disability and the Labor Market: Economic Problems, Policies, and Programs* (Ithaca, N.Y.: ILR Press, Cornell University, 1986).

CHAPTER 6: MEASURING AND DECIDING DISABILITY, *Donald O. Parsons*

1. For fine reviews of the historical evolution of the social security program, see Monroe Berkowitz, William G. Johnson, and Edward H. Murphy, *Public Policy toward Disability* (New York: Praeger Publishers, Inc., 1976), and Carolyn L. Weaver, "Social Security Disability Policy in the 1980s," in Monroe Berkowitz and M. Anne Hill, eds., *Disability and the Labor Market: Economic Problems, Policies, and Programs* (Ithaca, N.Y.: ILR Press, Cornell University, 1986), pp. 29–63.

2. It is important to distinguish here between conscious deception, which may be only a modest problem in the disability program, and self-deception, which is likely to be much more pervasive.

3. U.S. Department of Health and Human Services, Social Security Administration, *Social Security Handbook 1984* (Washington, D.C.: Government Printing Office, 1984), p. 75. See also Jong Chul Rhee, "Workers' Compensation as Income Insurance" (Ph.D. diss., Ohio State University, 1986), on how the limitation of coverage is a major concession to the disability measurement problem.

4. R. T. Smith and A. M. Lilienfeld, "The Social Security Disability Program: An Evaluation Study," U.S. Department of Health and Human Services, Social Security Administration, Office of Research and Statistics Research Report, no. 39 (Washington, D.C., 1971).

5. Saad Z. Nagi, *Disability and Rehabilitation: Legal, Clinical, and Self-Concepts and Measurement* (Columbus: Ohio State University Press, 1969).

6. See R. G. Dixon, Jr., *Social Security Disability and Mass Justice: A Problem in Welfare Adjudication* (New York: Praeger Publishers, Inc., 1973); Jerry L. Mashaw, Charles J. Goetz, Frank I. Goodman, Warren F. Schwartz, and Milton M. Carrow, *Social Security Hearings and Appeals* (Lexington, Mass.: Lexington Books, 1978); and Jerry L. Mashaw, *Bureaucratic Justice: Managing Social Security Disability Claims* (New Haven, Conn.: Yale University Press, 1983).

7. Dixon, *Social Security Disability.*

8. The situation has been somewhat different in the case of redeterminations of eligibility for people already on the rolls, which were conducted on a large-scale basis during the early 1980s. In the period of fiscal years 1980–1983, the average ALJ reversal rate in these cases was 60–65 percent. Beginning in 1982, moreover, benefits became payable during appeal through the ALJ hearing decision. For these reasons, it would appear that the appeals process was less random and the incentives for appeal stronger. The ALJ reversal rate on redeterminations now stands at 50 percent, after falling to a low of 32 percent in fiscal year 1985; benefits continue to be payable during appeal. For data, see U.S. Congress, Committee on Ways and Means, *Overview of Entitlement Programs: 1990 Green Book* (Washington, D.C.: U.S. Government Printing Office, June, 1990), p. 56.

9. For a theoretical analysis of the optimal disability program in a complete information environment, see Donald O. Parsons, "Social Insurance with Imperfect State Verification" (Ohio State University, 1988, Mimeographed).

10. For an analysis of the self-screening features of the current disability system, see Donald O. Parsons, "Self-Screening Mechanisms in a Targeted Public Transfer Program," *Journal of Political Economy*, forthcoming.

11. The labor market statistics in this paragraph are derived from data contained in U.S. Bureau of Labor Statistics, *Handbook of Labor Statistics*, Bulletin 2217, June 1985; the social security program statistics from U.S. Social Security Administration, *Social Security Bulletin: Annual Statistical Supplement, 1988* (Washington, D.C.: Government Printing Office, 1990).

12. Mordechai E. Lando, "The Effect of Unemployment on Application for Disability Insurance," *1974 Business and Economic Statistics Section, Proceedings of the American Statistical Association, 1974*. See also M. E. Lando, M. B. Coate, and Ruth Kraus, "Disability Benefit Applications and the Economy," *Social Security Bulletin*, no. 42 (October 1979), pp. 3–10.

13. See Donald O. Parsons, "The Decline in Male Labor Force Participation," *Journal of Political Economy*, vol. 88, no. 1 (February 1980), pp. 117–34; "Racial Trends in Male Labor Force Participation," *American Economic Review*, vol. 70, no. 5 (December 1980), pp. 911–20; and "The Male Labor Force Participation Decision: Health, Reported Health and Economic Incentives," *Economica*, vol. 49 (February 1982), pp. 81–91.

14. Jonathan S. Leonard, "Labor Supply Incentives and the Disincentives for Disabled Persons," in Monroe Berkowitz and M. Anne Hill, eds., *Disability and the Labor Market: Economic Problems, Policies, and Programs* (Ithaca, N.Y.: ILR Press, Cornell University, 1986).

15. For proposals to increase the importance of the substantial gainful activity level in the disability program, see U.S. Disability Advisory Council, *Report of the Disability Advisory Council* (Baltimore, Md.: Department of Health and Human Services, Social Security Administration, 1988).

16. Paula A. Franklin and John C. Hennessey, "Effect of Substantial Activity Level on Disabled Beneficiary Work Patterns," *Social Security Bulletin*, no. 42 (March 1979), pp. 3–10.

WEST GERMAN APPROACH, *Richard V. Burkhauser*

1. Richard V. Burkhauser and Petri Hirvonen, "United States Disability Policy in a Time of Crisis: A Comparison with Sweden and the Federal Republic of Germany," *Milbank Quarterly*, vol. 67, supp. 2 (1989), pp. 166–94.

2. Franz Brandt, *Ursachen für Schwierigkeiten bei der Eingliederung von Schwerbehinderten auf dem allgemeinen Arbeitsmarkt* (Why the severely disabled have difficulty integrating into the labor market) (Bonn: Bundesminister für Arbeit und Sozialordnung, 1985).

3. Specific laws mandating accommodation also exist in Europe. The Handicapped Workers' Employment Act was part of the Amendments of the Dutch Social Security Disability Act of 1987. Like the ADA it requires employees to accommodate the demands of a job and all other working conditions to the functional limitations of impaired employees. Unlike the ADA, however, employers that accommodate their workers in this way are entitled to compensation from the government. To date this has not been an important component of the Dutch disability system. In 1989 only 13 million guilders (about 7 million dollars) was spent on workplace accommodation

out of a total disability budget of 21.5 billion guilders. For a detailed discussion of the Dutch disability system, see Leo J. M. Aarts and Philip R. De Jong, *"Economic Aspects of Disability Behavior"* (Ph.D. diss., Erasmus University, 1990).

4. For detailed discussions of these issues, see Richard V. Burkhauser, "Morality on the Cheap: The Americans with Disability Act," *Regulation*, vol. 13, no. 2 (Spring 1990), and Carolyn L. Weaver, "The ADA: Another Mandated Benefits Program?" in the *American Enterprise*, vol. 1, no. 3 (May/June 1990), pp. 81–84.

5. Barry Bye and Evan Schechter, *1978 Survey of Disability and Work*, (Washington, D.C.: U.S. Department of Health and Human Services, Social Security Administration Publication, no. 13-11745, SSA, January 1982).

6. Richard V. Burkhauser and Yang Woo Kim, "The Importance of Employer Accommodation on the Job Duration of Health Impaired Workers: A Hazard Model Approach," Vanderbilt University Working Paper (1990).

Bibliography

Americans with Disabilities Act of 1990, 42 U.S. Code 12101 (July 26, 1990).

Andrews, Roxanne M., Martin Ruther, David K. Baugh, Penelope L. Pine, and Marilyn P. Rymer. "Medicaid Expenditures for the Disabled under a Work Incentive Program." *Health Care Financing Review* 9 (Spring 1988): 1–8.

Becker, Gary S. *The Economics of Discrimination.* 2d ed. Chicago: University of Chicago Press, 1971.

Bennefield, Robert L., and John M. McNeil. *Labor Force Status and Other Characteristics of Persons with a Work Disability: 1981–1988.* U.S. Department of Commerce, Bureau of the Census Series P-23, no. 160. Washington, D.C.: Bureau of the Census, July 1989.

Berkeley Planning Associates. *A Study of Accommodations Provided to Handicapped Employees by Federal Contractors.* Final Report to the U.S. Department of Labor, contract no. J-9-E-1-0009. 2 vols. Washington, D.C.: Government Printing Office, 1982.

Berkowitz, Monroe, David Dean, Dale Hanks, and Stanley Portny. *Enhanced Understanding of the Economics of Disability.* Final report to the Virginia Department of Rehabilitative Services. Richmond, VA, August 1988.

Berkowitz, Monroe, and Carolyn Green. "American Expenditures." *American Rehabilitation* (Spring 1989).

Berkowitz, Monroe, and M. Anne Hill, eds. *Disability in the Labor Market: Economic Problems, Policies, and Programs.* Ithaca, N.Y.: ILR Press, Cornell University, 1986.

Berkowitz, Monroe, William G. Johnson, and E. H. Murphy. *Public Policy toward Disability.* New York: Praeger Publishers, Inc., 1976.

Board of Trustees of the Federal Old-Age and Survivors Insurance and Disability Insurance Trust Funds. *1989 Annual Report.* Washington, D.C.: Government Printing Office, 1989.

Bound, John. "The Health and Earnings of Rejected Insurance Disability Insurance Applicants." *American Economic Review* 79 (June 1989): 482–503.

Bound, John, and Tim Waldman. "Disability Transfers and the Labor Force Attachment of Older Men: Evidence from the Historical Record." University of Michigan Working Paper. Ann Arbor: 1989.

Bowe, Frank. "Out of the Job Market: A National Crisis." Washington, D.C.: President's Committee on Employment of the Handicapped, 1986.

Brandt, Franz. *Ursachen für Schwierigkeiten bei der Eingliederung von Schwerbehinderten auf dem allgemeinen Arbeitsmarkt* (Why the severely disabled have difficulty integrating into the labor market). Bonn: Bundesminister für Arbeit und Sozialordnung, 1985.

Burgdorf, Robert L., Jr. *The Legal Rights of Handicapped Persons: Cases, Materials, and Text.* Baltimore, Md.: Paul H. Brookes, 1980.

Burkhauser, Richard V. "Morality on the Cheap: The Americans with Disabilities Act." *Regulation* 13, no. 2 (Summer 1990): 47–56.

———. "Disability Policy in the United States, Sweden, and the Netherlands." In *Disability and the Labor Market: Economic Problems, Policies, and Programs,* edited by Monroe Berkowitz and M. Anne Hill. Ithaca, N.Y.: ILR Press, Cornell University, 1986, pp. 262–84.

Burkhauser, Richard V., and Robert H. Haveman. *Disability and Work: The Economics of American Policy.* Baltimore, Md.: Johns Hopkins University Press, 1982.

Burkhauser, Richard V., and Petri Hirvonen. "United States Disability Policy in a Time of Crisis: A Comparison with Sweden and the Federal Republic of Germany." *Milbank Quarterly* 67, supp. 2 (1989): 166–94.

Burkhauser, Richard V., and Yang Woo Kim. "The Importance of Employer Accommodation on the Job Duration of Health Impaired Workers: A Hazard Model Approach." Vanderbilt University Working Paper. Nashville, Tenn.: Vanderbilt University, 1990.

Bye, Barry V., and Evan Schechter. *1978 Survey of Disability and Work: Technical Introduction.* U.S. Department of Health and Human Services, Social Security Administration Publication, no. 13-11745. Baltimore, Md.: SSA, January 1982.

Bye, Barry V., and Gerald F. Riley. "Eliminating the Medicare Waiting Period for Social Security Disabled Worker Beneficiaries." *Social Security Bulletin* 52, no. 5 (May 1989): 2–15.

Calabresi, Guido, and Alvin Klevorick. "Four Tests for Liability in Torts." *Journal of Legal Studies* 14, no. 585 (1985).

Calfee, John E., and Richard Craswell. "Some Effects of Uncertainty on Compliance with Legal Standards." *Virginia Law Review* 70, no. 965 (1984).

Collignon, Frederick C. "The Role of Reasonable Accommodation in Employing Disabled Persons in Private Industry." In *Disability and the Labor Market: Economic Problems, Policies, and Programs,* edited by Monroe Berkowitz and M. Anne Hill. Ithaca, N.Y.: ILR Press, Cornell University, 1986.

Cotterman, Robert F., and John Raisian. "The Incidence of Disability, 1970 to 2020: A Public Policy Dilemma?" A report submitted to the

National Council on the Disabled. Washington, D.C.: Unicon Research Corp., May 1988.

Dale, Charles V. "Remedies and Standing to Sue under Section 933, the 'Americans with Disabilities Act of 1989.'" CRS report for Congress. Washington, D.C.: U.S. Congressional Research Service, May 26, 1989.

Dejong, G., A. I. Batavia, and R. Griss. "America's Neglected Health Minority: Working-Age Persons with Disabilities." *Milbank Quarterly* 67, no. 2, supp. (1989): 311–51.

Disability Advisory Committee. *Report of Disability Advisory Committee to the Commissioner of Social Security*. Baltimore, Md.: U.S. Department of Health and Human Services, Social Security Administration, June 25, 1989.

Disability Advisory Council. *Report of the Disability Advisory Council*. Baltimore, Md.: U.S. Department of Health and Human Services, Social Security Administration, 1988.

Dixon, Robert G., Jr. *Social Security Disability and Mass Justice: A Problem in Welfare Adjudication*. New York: Praeger Publishers, 1973.

Franklin, Paula A. "Impact of Disability on the Family Structure." *Social Security Bulletin* 40, no. 5 (May 1977): 3–18.

Franklin, Paula A., and John C. Hennessey. "Effect of Substantial Activity Level on Disabled Beneficiary Work Patterns." *Social Security Bulletin* 42 (March 1979): 3–17.

Grossman, Michael. "The Concept of Health Capital and the Demand for Health." *Journal of Political Economy* 80 (March 1972): 223–55.

Haber, L. "Identifying the Disabled: Concepts and Methods in the Measurement of Disability." *Social Security Bulletin* 30 (December 1967): 17–35.

"Handicapped Workers: Who Should Bear the Burden of Proving Job Qualifications?" *Maine Law Review* 38, no. 135: 150 n.66.

Haveman, Robert H., and Barbara L. Wolfe. "Disability Transfers and Early Retirement: A Causal Relationship." *Journal of Public Economics* 22 (June 1984): 47–66.

———. "The Economic Well-being of the Disabled, 1962 to 1984." *Journal of Human Resources* (Winter 1990).

Haveman, Robert H., Barbara L. Wolfe, and Jennifer Warlick. "Disability Transfers, Early Retirement, and Retrenchment." In *Retirement and Economic Behavior*, edited by Henry Aaron and Gary Burtless, 65–93. Washington, D.C.: Brookings Institution, 1984.

Hayek, F. A. "The Use of Knowledge in Society." *American Economic Review* 35, no. 4 (September 1945): 519–30.

Hirvenon, Petri, and Richard V. Burkhauser. "German Disability in a Time of Crisis." Vanderbilt University Working Paper. Nashville, Tenn.: Vanderbilt University, May 1989.

Jones, Nancy Lee. "The Americans with Disabilities Act (ADA): An

Overview of Selected Major Legal Issues." U.S. Congressional Research Service report for Congress. Washington, D.C.: CRS, July 25, 1989.

"Judicial Review of Social Security Disability Decisions: A Proposal for Change." *Texas Law Review* 11, no. 215 (1980).

Kaplow, Louis. "Optional Government Policy toward Risk Imposed by Uncertainty Concerning Future Government Action." Harvard Program in Law and Economics, Discussion Paper, no. 20. Harvard University, August 1986.

Kosters, Marvin H. "Wages and Demographics." In *Workers and Their Changing Patterns in the United States,* edited by Marvin H. Kosters. Washington, D.C.: American Enterprise Institute, forthcoming.

Landis, William M. "The Economics of Fair Employment Laws." *Journal of Political Economy* 76 (July/August 1968): 507–45.

Lando, Mordechai E. "The Effect of Unemployment on Application for Disability Insurance." *1974 Business and Economic Statistics Section, Proceedings of the American Statistical Association.* 1974.

Lando, Mordechai E., Malcolm B. Coate, and Ruth Kraus. "Disability Benefit Applications and the Economy." *Social Security Bulletin* 42 (October 1979): 3–10.

Leonard, Jonathan S. "On the Decline in the Labor Force Participation Rates of Older Black Males." Honors thesis, Harvard University, 1976.

———. "The Social Security Disability Insurance Program and Labor Force Participation." National Bureau of Economic Research Working Paper, no. 392, Cambridge, Mass.: 1979.

———. "Labor Supply Incentives and Disincentives for the Disabled." In *Disability and the Labor Market: Economic Problems, Policies, and Programs,* edited by Monroe Berkowitz and M. Anne Hill, 64–96. Ithaca, N.Y.: ILR Press, Cornell University, 1986.

———. "Affirmative Action in the 1980s: With a Whimper Not a Bang." Berkeley: University of California, 1987.

———. "The Effectiveness of Equal Employment Law and Affirmative Action Regulation." *Research in Labor Economics* 8, part B (1987): 319–50.

Louis Harris and Associates, Inc. *The ICD Survey of Disabled Americans: Bringing Disabled Americans into the Mainstream.* New York: International Center for the Disabled, March 1986.

Mashaw, Jerry L. *Bureaucratic Justice: Managing Social Security Disability Claims.* New Haven, Conn.: Yale University Press, 1983.

Mashaw, Jerry L., Charles J. Goetz, Frank I. Goodman, Warren F. Schwartz, and Milton M. Carrow. *Social Security Hearings and Appeals.* Lexington, Mass.: Lexington Books, 1978.

McCoy, John L., and Kerry Weems. "Disabled Worker Beneficiaries and Disabled SSI Recipients: A Profile of Demographic and Program Characteristics." *Social Security Bulletin* 52 (May 1989): 16–28.

Murray, Charles. *Losing Ground, American Social Policy 1950 to 1980.* New York: Basic Books, 1984.

Nagi, Saad Z. *Disability and Rehabilitation: Legal, Clinical, and Self-Concepts and Measurement.* Columbus: Ohio State University Press, 1969.

National Council on the Handicapped. *On the Threshold of Independence.* A report to the president and the Congress from the National Council on the Handicapped. Washington, D.C.: NCH, January 1988.

———. *Toward Independence.* A report to the president and the Congress from the National Council on the Handicapped. Washington, D.C.: NCH, 1986.

Oi, Walter Y. "Three Paths from Disability to Poverty." U.S. Department of Labor, Office of the Assistant Secretary for Policy Evaluation and Research, Technical Analysis Paper, no. 58. Washington, D.C.: Government Printing Office, October 1978.

Parsons, Donald O. "The Decline in Male Labor Force Participation." *Journal of Political Economy* 88, no. 1 (February 1980): 117–34.

———. "Racial Trends in Male Labor Force Participation." *American Economic Review* (December 1980): 911–20.

———. "The Male Labor Force Participation Decision: Health, Reported Health and Economic Incentives." *Economica* 49 (February 1982): 81–91.

———. "Social Insurance with Imperfect State Verification: Income Insurance for the Disabled." Rev. ed. Columbus: Ohio State University, 1988. Mimeo.

———. "Self-Screening Mechanisms in a Targeted Public Transfer Program." Columbus: Ohio State University, 1989. Mimeo.

President of the United States. *Economic Report of the President.* Washington, D.C.: U.S. Government Printing Office, February 1990.

Priest, George L., and William Klein. "The Selection of Disputes for Trial." *Journal of Legal Studies* 13, no. 1 (1984).

Rhee, Jong Chul. "Workers' Compensation as Income Insurance." Ph.D. diss., Ohio State University, 1986.

School Board of Nassau County v. Arline, 107 Sup. Ct. 1123, 1987.

Smith, Mary F. "Section 504 of the Rehabilitation Act: Statutory Provisions, Legislative History, and Regulatory Requirements." U.S. Congressional Research Service report for Congress. Washington, D.C.: CRS, January 19, 1989.

Smith, Richard T., and Abraham M. Lilienfield. "The Social Security Disability Program: An Evaluation Study." U.S. Department of Health and Human Services, Social Security Administration Research Project, no. 39. Baltimore, Md.: SSA, 1971.

"Social Security Disability Amendments of 1980: Legislative History and Summary of Provisions." *Social Security Bulletin* 44, no. 4 (April 1981): 3–23.

Torres v. Bolger, 610 F. Supp. 593 D.C. Tex. 1985.

Tr   Ralph. "Recovery of Disabled Beneficiaries: A 1975 Followup Study of 1972 Allowances." *Social Security Bulletin* 42, no. 4 (April 1979): 3–23.

U.S. Congress. House. *Conference Report on the Disability Amendments of 1980*. 96th Congress, 2d. session, 1980. Rept. 96-944.

U.S. Congress. House. Committee on Ways and Means. *Background Material and Data on Programs within the Jurisdiction of the Committee on Ways and Means*. 101st Congress, 1st session, March 15, 1989. Committee Print 101-4.

———. *Overview of Entitlement Programs: 1990 Green Book*, 101st Congress, 2d session, June 5, 1990. Committee Print 101-29.

U.S. Congress. Senate. Committee on Finance. *Staff Data and Materials Related to the Social Security Disability Insurance Program*. 97th Congress, 2d session, August 1982. Committee Print 97-16.

U.S. Department of Health, Education, and Welfare, Rehabilitation Services Administration. "The Comprehensive Needs Study." Washington, D.C.: Urban Institute, 1975.

U.S. Department of Health and Human Services. Social Security Administration. Office of Legislative and Regulatory Policy. *Social Security Handbook: 1984*. Washington, D.C.: Government Printing Office, 1984.

———. *Social Security Bulletin: Annual Statistical Supplement, 1988*. Washington, D.C.: Government Printing Office, 1989.

U.S. Department of Labor. Bureau of Labor Statistics. *Handbook of Labor Statistics*. Bulletin 2217. Washington, D.C.: Government Printing Office, June 1985.

Verbrugge, L. "Longer Life but Worsening Health: Trends in Health and Mortality of Middle-Aged and Older Persons." *Milbank Memorial Fund Quarterly* 62, no. 3 (1984): 475–519.

Veterans Readjustment Act of 1974, 38 U.S.Code 42.

Weaver, Carolyn L. "Social Security Disability Policy in the 1980s and Beyond." In *Disability and the Labor Market: Economic Problems, Policies, and Programs*, edited by Monroe Berkowitz and M. Anne Hill. Ithaca, N.Y.: ILR Press, Cornell University, 1986, pp. 29–63.

———. "The ADA: Another Mandated Benefits Program?" *American Enterprise* 1, no. 3 (May/June 1990): 81–84.

A Note on the Book

This book was edited by Ann Petty and Cheryl Weissman
of the staff of the AEI Press.
The text was set in Palatino, a typeface designed by the twentieth-century
Swiss designer Hermann Zapf. Coghill Composition Company,
of Richmond, Virginia, set the type, and
Edwards Brothers Incorporated, of Ann Arbor, Michigan,
printed and bound the book, using permanent acid-free paper.

The AEI Press is the publisher for the American Enterprise Institute for Public Policy Research, 1150 17th Street, N.W., Washington, D.C. 20036: *Christopher C. DeMuth*, publisher; *Edward Styles*, director; *Dana Lane*, editor; *Ann Petty*, editor; *Cheryl Weissman*, editor; *Susan Moran*, editorial assistant (rights and permissions).

Books published by the AEI Press are distributed by arrangement with the University Press of America, 4720 Boston Way, Lanham, Md. 20706.